Heike Lueger

The variability of the surface pCO2 in the North Atlantic Ocean

Heike Lueger

The variability of the surface pCO2 in the North Atlantic Ocean

Results from continuous CO2 measurements onboard the car carrier MS Falstaff

Südwestdeutscher Verlag für Hochschulschriften

Impressum/Imprint (nur für Deutschland/ only for Germany)
Bibliografische Information der Deutschen Nationalbibliothek: Die Deutsche Nationalbibliothek verzeichnet diese Publikation in der Deutschen Nationalbibliografie; detaillierte bibliografische Daten sind im Internet über http://dnb.d-nb.de abrufbar.

Alle in diesem Buch genannten Marken und Produktnamen unterliegen warenzeichen-, marken- oder patentrechtlichem Schutz bzw. sind Warenzeichen oder eingetragene Warenzeichen der jeweiligen Inhaber. Die Wiedergabe von Marken, Produktnamen, Gebrauchsnamen, Handelsnamen, Warenbezeichnungen u.s.w. in diesem Werk berechtigt auch ohne besondere Kennzeichnung nicht zu der Annahme, dass solche Namen im Sinne der Warenzeichen- und Markenschutzgesetzgebung als frei zu betrachten wären und daher von jedermann benutzt werden dürften.

Verlag: Südwestdeutscher Verlag für Hochschulschriften Aktiengesellschaft & Co. KG
Dudweiler Landstr. 99, 66123 Saarbrücken, Deutschland
Telefon +49 681 37 20 271-1, Telefax +49 681 37 20 271-0
Email: info@svh-verlag.de
Zugl.: Kiel, Christian-Albrechts-Universitaet, Diss., 2004

Herstellung in Deutschland:
Schaltungsdienst Lange o.H.G., Berlin
Books on Demand GmbH, Norderstedt
Reha GmbH, Saarbrücken
Amazon Distribution GmbH, Leipzig
ISBN: 978-3-8381-0865-0

Imprint (only for USA, GB)
Bibliographic information published by the Deutsche Nationalbibliothek: The Deutsche Nationalbibliothek lists this publication in the Deutsche Nationalbibliografie; detailed bibliographic data are available in the Internet at http://dnb.d-nb.de.

Any brand names and product names mentioned in this book are subject to trademark, brand or patent protection and are trademarks or registered trademarks of their respective holders. The use of brand names, product names, common names, trade names, product descriptions etc. even without a particular marking in this works is in no way to be construed to mean that such names may be regarded as unrestricted in respect of trademark and brand protection legislation and could thus be used by anyone.

Publisher: Südwestdeutscher Verlag für Hochschulschriften Aktiengesellschaft & Co. KG
Dudweiler Landstr. 99, 66123 Saarbrücken, Germany
Phone +49 681 37 20 271-1, Fax +49 681 37 20 271-0
Email: info@svh-verlag.de

Printed in the U.S.A.
Printed in the U.K. by (see last page)
ISBN: 978-3-8381-0865-0

Copyright © 2010 by the author and Südwestdeutscher Verlag für Hochschulschriften Aktiengesellschaft & Co. KG and licensors
All rights reserved. Saarbrücken 2010

The variability of surface pCO$_2$ in the North Atlantic Ocean by Heike Lueger

Table of Contents:

1.	**Introduction**		1
2.	**Scientific Background**		2
	2.1. Carbon Dioxide Chemistry of the Sea		2
	2.1.1. The Carbonate System		2
	2.1.2. Time Dependence of pCO$_2$		5
	2.1.3 Air-Sea Flux of CO$_2$		6
	2.1.4. Exchange with the Organic Carbon Pool		7
3.	**Long-term Monitoring of CO$_2$ in the North Atlantic**		9
	3.1. Carbon Variability Studies by Ships of Opportunity		10
	3.1.1. pCO$_2$ Measurements onboard M/V Falstaff		12
4.	**Major Conclusions and Outlook**		13
5.	**References**		14
6.	**Chapter I:**	The pCO$_2$ Variability in the North Atlantic Ocean	17
7.	**Chapter II:**	The Seasonality of the Air-Sea CO2 Flux in the Mid-Latitude North Atlantic	44
8.	**Chapter III:**	Seasonal Cycles of Nutrients in the North Atlantic	68

The variability of surface pCO$_2$ in the North Atlantic Ocean by Heike Lueger

The variability of surface pCO$_2$ in the North Atlantic Ocean by Heike Lueger

1. **Introduction**

The global climate change is inseparately connected to the carbon cycle with carbon dioxide as one of the major greenhouse gases. It is a well known fact that anthropogenic emissions have led to a higher atmospheric content of CO$_2$ since the onset of the industrial revolution. Such a high atmospheric CO$_2$ concentration is unprecedented and has not been exceeded during the past 420,000 years (Houghton et al., 2001). The atmosphere, however, only stores about half of the anthropogenic CO$_2$ and it remains unclear what happens to the other half. Generally, the atmosphere exchanges CO$_2$ in a source/ sink pattern with two major reservoirs: the terrestrial biosphere and the ocean. The atmospheric CO$_2$ content is well known for two reasons. Firstly, in the atmosphere CO$_2$ is distributed uniformly due to the rapid mixing and, secondly, a high quality world-wide network exists which continously monitors the atmospheric CO$_2$ content (Conway et al., 1994). On land and in the ocean the CO$_2$ variability and consequently carbon storage is much more difficult to determine. A vast multitude of carbon species exists in the terrestrial biosphere and land use is continuously changing which makes it very difficult to constrain carbon storage (Wallace, 2001). In the ocean most of the carbon is present as inorganic carbon and the carbon uptake term can be estimated with higher reliability than in the terrestrial biosphere. Thus the latter is commonly calculated from the difference between atmospheric and oceanic reservoir. A lot of research has been conducted to find out about the general mechanisms that underlie the complex carbon cycle in the ocean. The North Atlantic Ocean plays an important role in the global ocean with regard to the uptake of anthropogenic CO$_2$. Deep-water formation in the high northern latitudes of the Atlantic leads to a deep penetration of anthropogenic CO$_2$ (Gruber, 1998).

Based on this scientific knowledge the European project CAVASSOO (Carbon Variability Studies by Ships of Opportunity) was launched as a pilot project in 2000. Within my thesis - which was part of this project - I came across major questions and topics that are crucial for a better understanding of the carbon cycling and global implications such as climate change:

- How does the marine carbon cycle work? What are the patterns of CO$_2$ within the ocean and at the surface?
- How variable is the CO$_2$ flux and related parameters in the North Atlantic Ocean?
- How well is the international database established? How important is the addition of new CO$_2$ data and parameters such as e.g. wintertime nutrients?

The variability of surface pCO$_2$ in the North Atlantic Ocean by Heike Lueger

2. Scientific Background

2.1. Carbon dioxide chemistry of the sea

2.1.1. The Carbonate System

The carbon cycle of the ocean is well constrained with regards to pathways of CO_2, however, actual fluxes and quantities are an ongoing subject of debate. CO_2 is exchanged between atmosphere and ocean, where the flux largely depends on the concentration gradient. In ocean

---

Box 1: Reaction Chemistry of CO_2 in seawater.

CO_2 from the atmosphere dissolves in seawater:

(1) $CO_2 (g) \overset{K_H}{\Leftrightarrow} CO_2 (f)$ $\quad K_H$: Henry constant (solubility constant)
    (g: gas phase; f: free uncombined $CO_2 = pCO_2$)

(2) $CO_2 (f) + H_2O \Leftrightarrow H_2CO_3$ SLOW REACTION

$H_2CO_3{}^*$ or $CO_2(aq)$ summary of equation (2); denotes the non-ionized dissolved CO_2

complete reaction with subsequent dissolution:

$CO_2 (f) + H_2O \overset{K}{\Leftrightarrow} H_2CO_3 \overset{K_1}{\Leftrightarrow} H^+ + HCO_3^- \overset{K_2}{\Leftrightarrow} H^+ + CO_3^{2-}$

$\quad\quad\quad\quad\quad\quad\quad\quad$ carbonic acid $\quad\quad$ bicarbonate $\quad\quad$ carbonate

K, K_1, K_2: dissociation constants

---

areas where the partial pressure of CO_2 is higher in the ocean than in the atmosphere the flux will be from the ocean into the atmosphere – this pattern is called outgassing, i.e. the ocean acts as a source of CO_2 for the atmosphere. Hence, the reverse process denotes the ocean to act as a sink for CO_2 to the atmosphere. The air-sea exchange of CO_2 is difficult to measure directly and is controlled by myriad processes, e.g.: wind speed, sea state, surface processes, bubble entrainment, bioproductivity (McGillis et al., 2001). The CO_2 flux is mostly derived from estimated transfer velocities, solubility of CO_2, and the difference between the partial pressure of CO_2 in the bulk seawater and at the ocean surface (ΔpCO_2). This procedure is a controversial subject largely because the transfer velocity is estimated from the wind speed and this parameterization is not well constrained as of yet (e.g. Wanninkhof, 1992; Wanninkhof and McGillis, 1999; Nightingale et al., 2000).

Box 2: Parameters of the CO_2 system.

Parameter	accuracy[1]	chemical determination[2]	typical seawater value
pH	± 0.002	spectrophotometric (Friis et al., in press)	8.0 - 8.2
DIC[3]	± 2 µmol/kg	coulometric (Johnson et al., 1999)	2100 µmol/kg
TA	± 4 µmol/kg	potentiometric (Mintrop et al., 2000)	2300 µmol/kg
pCO_2	± 1 µatm	non-dispersive infrared (Murphy et al., 2001)	350 µatm

[1] achievable accuracy by state-of-the-art measurements
[2] there is more than one possible detection, mentioned are only the most common techniques
[3] sum of inorganic carbon species, also known as TC or CT (total carbon)

Once CO_2 gas enters the ocean it reacts – unlike inert gases such as oxygen or nitrogen - with water and subsequently dissolves (box 1). The hydration of CO_2 to carbonic acid is a slow reaction: In one second, only about 3 % of the molecules of free CO_2 in the solution combine with a water molecule ($K_H \approx 0.03$ s^{-1}; cf. Pilson, 1998). Equation 2 (box 1) is generally summarized into the term [CO_2*] (or [CO_2 aq]) where carbonic acid is present in a ratio of [CO_2 fl] : [H_2CO_3] = 650 : 1 (at 25°C). The dissolution of carbon dioxide into bicarbonate and carbonate is controlled by dissociation constants which depend on pressure, temperature and salinity. Often they are referred to as apparent dissociation constants (Mehrbach et al., 1973). The complete CO_2 chemistry of the sea which involves the reactions, CO_2 species and constants is also called CO_2 or carbonate system. It is not possible to analytically detect the concentrations of carbonic acid, bicarbonate or carbonate. Nevertheless the entire CO_2 system can be calculated from the combination of any two of the four parameters: pH, pCO_2 (partial pressure of CO_2), DIC (dissolved inorganic carbon), and TA (total alkalinity). All these parameters can be measured by chemical analysis with different accuracy (box 2).

The negative logarithm of the proton concentration, better known as pH, is considered as the 'master variable' of the carbonate system as it determines the speciation of CO_2. Seawater has a slightly basic pH (~ 8), therefore the speciation of the carbonate parameters approximates the following ratios: 100 (HCO_3^-) : 10 (CO_3^{2-}) : 1 ($H_2CO_3^*$).

The partial pressure of CO_2 (pCO_2) is calculated from the total pressure (p) and the mole fraction of CO_2 (xCO_2):

$$pCO_2 = p \cdot xCO_2 \qquad (1)$$

The common detection method involves a non-dispersive infrared technique which can be used in a continuous measurement mode. Another method for the determination of pCO_2 is a gaschromatographic measurement technique which is less rugged and commonly used for discrete samples.

Total dissolved inorganic carbon (DIC) encompasses the sum of the concentration of the CO_2 species:

$$DIC = [H_2CO_3^*] + [HCO_3^-] + [CO_3^{2-}] \qquad (2)$$

The coulometric technique for the determination of DIC involves the acidification of a sample of seawater which leads to an outgassing of CO_2. The CO_2 gas sample is titrated and the CO_2 concentration is detected coulometrically.

Total Alkalinity (TA), the fourth parameter of the carbonate system, describes the buffer capacity of for example seawater and encompasses bases that are formed from weak acids. It is determined by titrating a sample with strong acid (HCl) to the point where all such bases, mainly carbonate species, are protonated (cf. Pilson, 1998). Generally, TA is described as following:

$$TA = 2 [CO_3^{2-}] + [HCO_3^-] + [B(OH)_4^-] + [OH^-] - [H^+] \qquad (3)$$

Total alkalinity is a conservative quantity and linearly correlated with salinity. The carbonate species and to a lesser extent borate contribute mostly to the total alkalinity, but clearly every similar acid-base system present in solution will influence TA. In a typical seawater sample (TA = 2300 µmol/kg, DIC = 2000 µmol/kg, T = 25°C, Salinity = 35) the carbonate alkalinity contributes 96% to the total alkalinity (cf. Zeebe and Wolf-Gladrow, 2001).

2.1.2. Time dependence of pCO_2

The atmospheric CO_2 content increases steadily due to human activities; the ocean is affected by this and follows the increase. The driving force that leads to a source/sink pattern within the ocean is the concentration gradient between atmospheric and oceanic pCO_2. The distribution of the surface pCO_2 in the ocean is much less homogenous than in the atmosphere as it also reflects biological, physical and chemical reaction schemes and less rapid mixing. With the onset of the industrial revolution the atmospheric pCO_2 of 280 µatm has increased to a recent value of circa 370 µatm in more than a century. The implication for the oceanic pCO_2 is significant as the driving force of the concentration gradient has changed signs.

The regional variability of the seawater pCO_2 is considerable. Physical processes, thermodynamical and biological processes affect the carbonate system. It is important to gather enough pCO_2 data to resolve the seasonal variability of the oceanic pCO_2 and subsequently of the CO_2 flux between the atmosphere and the ocean. The most prominent pCO_2 database is undoubtly the climatology by Takahashi et al. who published their most recent version in 2002. About 940,000 datapoints of the global oceans were collected and interpolated to a 4°x5° grid. They not only published pCO_2 data but also calculated the CO_2 flux and referenced all data to the year 1995 which was a non-El Nino year.

A major problem in this context is the strong time-dependency of the seawater pCO_2. It is a fact that the ocean takes up CO_2, but the knowledge does not extent to where and when CO_2 is taken up. Generally an oceanic increase in CO_2 on the order of 1.5 parts per million (ppm) and per year is assumed following the atmospheric increase. Takahashi et al. (2002) claim that this correction factor should not be applied to surface waters in high latitude regions (>45°N) since here the deep water formation takes place which dilutes the anthropogenic CO_2 signal. This assumption was derived from observations at Weather Station "P" (50°N, 145°W) in the northeastern subarctic Pacific by Wong and Chan (1991) where it was shown that the oceanic pCO_2 did not follow the atmospheric increase of CO_2. Takahashi et al. (2002) applied the pattern to all latitudes >45°N, which, however, may not be valid. Observations (Omar et al., in press) and models (e.g. Anderson and Olsen, 2002) show that the pCO_2 within this region did indeed increase, therefore it seems reasonable to apply the correction factor also to regions north of 45°N in the North Atlantic.

2.1.3. Air-Sea Flux of CO_2

The air-sea gas exchange of CO_2 plays an important role for the carbon chemistry in surface waters. The net flux of CO_2 ($F(CO_2)$) between the atmosphere and the surface ocean is difficult to measure directly and is commonly expressed by the following bulk formula:

$$F (CO_2) = k\, K_0\, (pCO_{2\,sw} - pCO_{2\,atm}) \qquad (1)$$

where k is the gas transfer velocity, K_0 is the solubility of CO_2 in seawater, pCO_{2sw} and pCO_{2atm} are the partial pressures of CO_2 in seawater and atmosphere, respectively. The transfer velocity is commonly related to wind speed (Liss and Merlivat, 1986; Wanninkhof, 1992; Wanninkhof and McGillis, 1999). But there is still controversy about the best relationship as wind is not the only factor controlling the gas exchange. Gas transfer rates can also be affected by waves, bubbles and surfactants. However, wind speed can be measured fairly easily and is linked to most of these features, so it seems reasonable to relate k and wind speed. Much effort has been raised to find a universal relationship between the two. Liss and Merlivat (1986; hereafter referred to as LM86) proposed a parameterization based on three different wind speed regimes. They assume a smooth surface regime when wind speed is low, a rough surface regime for medium to higher wind speed and a breaking wave or bubble regime for very high wind speeds. Hence they propose the following relationships between k and the wind speed (u):

$k = 0.17\, u\, (Sc / 600)^{-2/3}$ $\qquad$ for u < 3.6 m/s $\qquad$ (2a)

$k = 2.85\, u - 9.65\, (Sc/600)$ $\qquad$ for 3.6 < u < 13 m/s $\qquad$ (2b)

$k = 5.9\, u - 49.3\, (Sc / 600)^{-1/2}$ $\qquad$ for u > 13 m/s $\qquad$ (2c)

where Sc is the Schmidt number which is temperature dependent and accounts for the viscosity of the medium and diffusivity of CO_2. A Schmidt number of 600 corresponds to CO_2 in freshwater at 20°C. This approach is based on wind tunnel experiments and lake experiments where they used the tracer gas SF_6. It is noteworthy to mention that in the lake experiment wind speeds did not exceed 8 m s^{-1} therefore LM86 extrapolated the results to higher wind speeds.

The quadratic parameterization of k by Wanninkhof (1992; hereafter referred to as W92) shows a stronger dependence on wind speed than LM86. It also considers the source of the wind speed data, i.e. whether they are short-term or climatological. The relationships between k and u which is calibrated with bomb and natural ^{14}C invasion rates is as follows:

$k = 0.39\, u^2\, (Sc/660)^{-1/2}$ $\qquad$ (climatological winds) $\qquad$ (3a)

$k = 0.31\, u^2\, (Sc/660)^{-1/2}$ $\qquad$ (short term winds) $\qquad$ (3b) where the Schmidt number of 660 is valid for seawater at 20°C.

A more recent parameterization of k is suggested by Wanninkhof and McGillis (1999; hereafter referred to as WM99). WM99 propose a cubic relationship between k and short-term wind speed which they also extrapolate to long-term based wind speeds based on ^{14}C experiments. This approach resulted from direct measurements of the CO_2 fluxes during a cruise in the North Atlantic (McGillis, 1999) and the dependence between k and u shows two patterns. The stronger dependence of k at higher wind speeds is explained by bubble entrainment, whereas the lower dependence of k at lower wind speed is attributed to retardation of the gas transfer by surfactants. Thus WM99 proposed the following relationship:

$$k = 0.0283 \; u^3 \; (Sc/660)^{-1/2} \qquad (4)$$

Another recent approach by Nightingale (2000; N00) uses a quadratic parameterization based on a conservative tracer study. The relationship between the gas transfer velocity and wind speed shows a dependence intermediate between that of Liss and Merlivat (1986) and Wanninkhof (1992):

$$k = 0.222 \; u^2 + 0.333 \; u \; (Sc/600)^{-1/2} \qquad (5)$$

N00 used microbial non-volatile tracers – bacterium spores – in order to derive the transfer velocity of volatile tracers ($^3He/SF_6$) and to estimate the Schmidt number.

2.1.4. Exchange with the Organic Carbon Pool

There are three 'reactive' zones in which the inorganic carbon concentration is changed by biological reactions: surface, water column, and sediment.

In this work I am concerned with the surface zone which denotes the upper part of the ocean where water masses are still affected by wind, sunlight and weather conditions. The depth at which 1 % of the daylight is still visible is referred to as the euphotic zone. Another zone of importance is the mixed layer depth. This is the depth where vertical mixing leads to a more-or-less homogenous distribution of temperature, salinity and nutrients. The depth of the mixed layer varies seasonally with deeper depths during the winter months when convection increases. In the northern North Atlantic for example one can observe mixed layer depths more than 500 m. At the surface the effects of CO_2 air-sea exchange and biological reactions such as photosynthesis, respiration and calcification are significant. We already introduced the effect of air-sea exchange. Photosynthesis consumes inorganic carbon and produces particulate organic matter and the reverse reaction corresponds to respiration. The uptake of CO_2 during photosynthesis decreases the pCO_2 and increases the gradient of CO_2 between surface ocean and lower atmosphere (ΔpCO_2). The overall reaction can be displayed as following:

$$106\ CO_2 + 16\ NO_3^- + HPO_4^{2-} + 18\ H^+ + 122\ H_2O \Leftrightarrow (CH_2O)106(NH_3)16H_3PO_4 + 138\ O_2 \quad (1)$$

Equation 1 represents the model by Redfield et al. (1963); this so called Redfield ratio which encompasses the ratio of C:N:P = 106:16:1 is still a matter of debate. Different approaches have been used to improve these elemental ratios (e.g. Anderson and Sarmiento, 1994; Anderson, 1995; Körtzinger et al., 2002). The Redfield ratio is often used to estimate the new production. New production is defined as that part of the primary production that is based on allochthonous nutrients, i.e. nutrients newly introduced to the euphotic zone (Dugdale and Goering, 1967). This contrasts the regenerated production which is based on recycled nutrients. Particulate and dissolved matter form total organic matter that is produced by biological reaction chains of the primary production. A part of the organic matter sinks, e.g. as marine snow, and is consumed in the deeper parts of the ocean. It is estimated that roughly 10% of the organic matter is exported below the euphotic zone. The export of organic matter is part of the biological pump and plays an important role for the sequestration of carbon to the deep ocean. About 80% of the water column gradients in dissolved inorganic carbon are due to the biological pump (Sarmiento et al., 1995).

Excursus:

It could be shown that the consumption of DIC relative to nitrate may considerably exceed the Redfield C:N ratio (Sambrotto, 1993; Körtzinger et al., 2002). This is referred to as carbon overconsumption (Toggweiler, 1993) and has some implications. It is possible that new nitrogen is introduced not via nitrate but through N_2 fixation, for instance by cyanobacteria. These bacteria, a prominent example is Trichodesmium, can utilize dinitrogen as an alternative nitrogen source. Oceanic N_2 fixation may directly influence the sequestration of CO_2 by providing a source of new nitrogen in the euphotic zone. These observations suggest a re-assessment of the concept of new production (Karl et al., 2002).

Particulate organic carbon (POC) is not the only export mode of carbon into the deep ocean (organic carbon pump). It can also be exported via calcification, i.e. the formation of calcium carbonate (calcium carbonate pump). The process of precipitation and dissolution of calcium carbonate ($CaCO_3$) also modulates the carbon chemistry in the ocean. Marine organisms form calcareous shells from calcium and carbonate following this equation:

$$Ca^{2+} + 2\ HCO_3^- \leftrightarrow CaCO_3 + CO_2 + H_2O \quad (1)$$

This chemical equation is a summary of a more complex reaction. The formation of calcium carbonate increases the concentration of CO_2, but at the same time consumes 2 moles of bicarbonate (HCO_3^-). This counterintuitive behaviour reduces the total alkalinity (TA) and dissolved inorganic carbon (DIC) by 2 and 1 units, respectively, and increases the pCO_2 at the same time (Zeebe and Wolf-Gladrow, 2001). $CaCO_3$ in the ocean is present in two forms: aragonite and calcite. Aragonite which is orthorhombic is less stable and hence more soluble than calcite, therefore it is less abundant than calcite (Pilson, 1998). The major calcite producers in the ocean are coccolithophorids and foraminifera, while the most abundant pelagic aragonite organisms are pteropods (Zeebe and Wolf-Gladrow).

Excursus:
Calcifiying organisms in the ocean respond to the rising atmospheric CO_2 content (Riebesell, 2000). It could be shown that coccolithophorids suffer from the increasing atmospheric CO_2 due to decreasing calcification success. The fundamental difference between calcifying and non-calcifying organisms is their feedback on the air-sea exchange of CO_2. Non-calcifying phytoplankton, such as diatoms which produce siliceous shells, reduce the CO_2 concentration in the upper ocean by their photosynthetic activity. Calcifying organisms, however, next to their photosynthetical acitivity also produce CO_2 through calcification thus offsetting some of the CO_2 drawdown.

3. Long-term monitoring of CO_2 in the North Atlantic

Many parts of the global oceans are still undersampled with regard to CO_2 measurements and the oceanic uptake rate of CO_2 is commonly based on model calculations rather than on observations. These models agree well on the overall size of the oceanic CO_2 sink, but latitudinal distributions can differ very greatly (Wallace, 2001). The need to improve the current CO_2 database supports long-term measurements such as time-series and projects using volunteer observing ships. Time series in the ocean exist for more than 20 years, e.g. Bermuda Atlantic Time-Series (BATS), Hawaii Ocean Time-Series (HOT), and they yielded very significant contributions in many aspects including CO_2 measurements. More recently chemical oceanographers adopted the use of volunteer observing ships (VOS), which have long before been recognized as helpful collectors of meteorological or oceanographical data. For example, worldwide between 2000 and 3000 VOS are employed which report weather conditions within the World Weather Watch Programs that are used for weather predictions and planning of ship routes (Sy, 1993).

The variability of surface pCO$_2$ in the North Atlantic Ocean by Heike Lueger

3.1. Carbon Variability Studies by Ships of Opportunity (CAVASSOO)

Continuous CO$_2$ monitoring projects employing measurements of the partial pressure of CO$_2$ in seawater onboard VOS have been attempted since the 1990's (e.g. Dandonneau, 1995; Cooper et al., 1998), but they were mostly short-term, i.e. less than one year. In 1995 the commercial vessel M/S Skaugran, a lumber carrier that sailed the North Pacific, was equipped with an autonomously operating pCO$_2$ unit supported by a bilateral collaboration between the National Institute of Environmental Studies (NIES) in Japan and the Institute of Ocean Sciences (IOS) in Canada. The vessel operated for more than four years and compiled a huge dataset of CO$_2$ and related parameters. The program is still continuing, albeit with different vessels.

There was no comparable ocean pCO$_2$ monitoring program in the North Atlantic until the start of the EU-funded program CAVASSOO (Carbon Variability Studies by Ships of Opportunity). The project which supported this PhD project was launched end of 2000 by five European research institutes: Institute of Marine Research (IfM, Kiel, Germany), University of East Anglia (Norwich, United Kingdom), Geophysical Institute (University of Bergen, Norway), Instituto de Investigaçiones Mariñas (Vigo, Spain), Laboratoire de Sciences du Climat et de l'Environnement (Gif-sur-Yvette, France). The major goal of this project was to assemble a large amount of seawater pCO$_2$ data from an ocean-wide network to estimate the CO$_2$ flux and its variability. Furthermore it was planned to increase the collaboration between scientific communities studying marine aspects of the carbon cycle and those working on carbon modelling, e.g. inverse modelling, and atmospheric measurements.

The pCO$_2$ network consisted of four shipping routes (Figure 1). The M/V Nuka Arctica is a cargo ship that belongs to the Royal Arctic Line and sails between Aalborg, Denmark, and Nuuk, Greenland, on a monthly basis. This VOS - operated by the Norwegian partner – was equipped with a Submersible Autonomous Moored Instrument (SAMI) which measures pCO$_2$ in a flow through mode and a thermosalinograph to monitor temperature and salinity. The M/V Santa Lucia is operated by the UK partner and it is a reefer that carries bananas from the Caribbean (Windward Islands) to Portsmouth, UK. The VOS was equipped with a pCO$_2$

The variability of surface pCO₂ in the North Atlantic Ocean by Heike Lueger

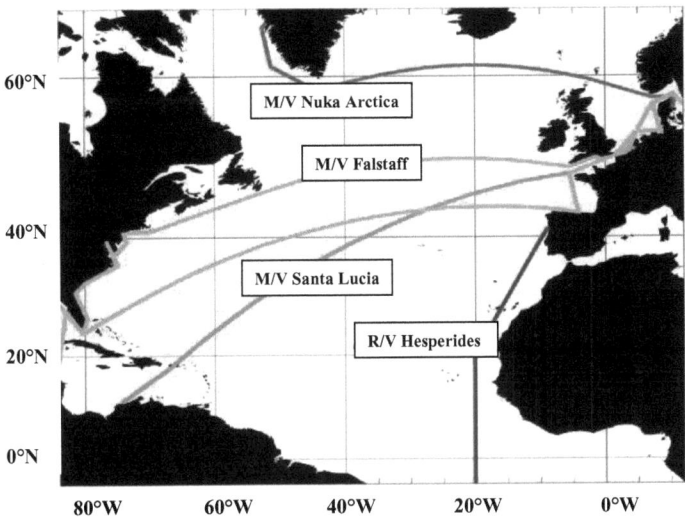

Figure 1: Volunteer observing ships and their shipping routes that continuously monitor pCO_2 in the North Atlantic Ocean. M/V Nuka Arctica: VOS line maintained by the university of Bergen, Norway. M/V Falstaff: VOS line maintained by the Institute for Marine Research, Kiel, Germany. M/V Santa Lucia: VOS line operated by the University of East Anglia, Norwich, United Kingdom. R/V Hesperides: Research vessel operatated by the Instituto de Investigaciones Marinas, Vigo, Spain.

measurement unit built at the UEA that includes a thermosalinograph and continuously monitors pCO_2. The R/V Hesperides is an Antarctic supply vessel which sails twice a year between Spain and the Antarctica. She is equipped with a thermosalinograph, fluorescence sensor and pCO_2 unit. At our institute (IfM, Kiel) we maintain the VOS line that includes the car carrier M/V Falstaff. Between February, 2002, and February, 2003, this car carrier - owned by the Swedish shipping company Wallenius – sailed between the UK and the East Coast of North America.

The variability of surface pCO$_2$ in the North Atlantic Ocean by Heike Lueger

3.1.1. *p*CO$_2$ Measurements on board M/V Falstaff

We equipped M/V Falstaff with a *p*CO$_2$ measurement unit beginning of 2002. The *p*CO$_2$ unit was generously lent to us by Yukihiro Nojiri (NIES, Japan) who was one of the main initiators of the VOS cruises in the North Pacific. This system is state-of-the-art and was built by Y.Nojiri and Kimoto Inc. (both Tsukuba, Japan). A detailed description of this system is given in Murphy et al. (2001).

The seawater intake is located at the starboard seachest of the vessel and is controlled and guarded by three safety valves. The seawater flows by hydrostatic pressure only as the engine room is located approx. 7 m below the waterline. A high-quality temperature sensor for measurement of the in-situ seawater temperature is installed next to the intake. From the seachest the seawater flows through an insulated hose to the thermosalinograph (model SBE21, Seabird Electronics Inc., Seattle/USA) where temperature and conductivity (for calculation of salinity) are continuously measured. The thermosalinograph is on free temporary loan from the Federal Maritime and Hydrographic Agency of Germany. The data are taken every six seconds and stored as 1-minute averages. The *p*CO$_2$ is also measured regularly in atmospheric air. For this purpose uncontaminated air is pumped from an air intake on the upper deck into the analysis system in the engine room through 100 m of flexible tubing. In addition to these measurements chlorophyll fluorescence is recorded continuously on a small bypass of the seawater flow (1-2 L/min) by a fluorescence sensor (Minitracka Mk II, Chelsea Technologies Group, West Molesey, Surrey, UK). All system operations and data handling are performed in an automated mode.

4. Major Conclusions and Outlook

- Our dataset shows that the western basin (35°W-70°W) displays a higher variability of seawater pCO$_2$ than the eastern basin (10°W-35°W). In the eastern basin the temperature and biology effect cancel each other and the overall effect is a subdued seasonal cycle of the seawater pCO$_2$. The western basin, however, is temperature controlled.

- We cannot estimate pCO$_2$ from chlorophyll. However, we found a promising correlation between mixed layer depth and pCO$_2$.

- This region of the North Atlantic ocean was a sink for CO$_2$ most of 2002. When comparing the CO$_2$ uptake rate in the region between 38°N and 46°N between our dataset and the climatology by Takahashi et al. (2002) using the same wind speed and grid resolution, the difference is small (4 %). However, our observation indicate that also data north of 45°N should be corrected for the increase in anthropogenic CO$_2$.

- During spring time we found a decrease in alkalinity concentrations for the NADR regime and showed that it could be attributed to calcification which increased the seasonal DIC loss by about 12%.

- The C:N ratio of the new production was highest for the GSTR (15.4) province accompanied by elevated N:P ratio (28.2) which indicates the strong influence of N$_2$ fixation in this region.

- We will continue the VOS line in cooperation with Wallenius within a future EU-project.

- A comparison between Atlantic and Pacific VOS data is planned and such is the exploitation of remaining discrete data (δ^{13}DIC, POC/DOC).

- We will also work out the physical effects on the pCO$_2$ within our dataset using a 1-dimensional box model.

5. References

Anderson, L.A., Olsen, A., 2002. Air-sea flux of anthropogenic carbon dioxde in the North Atlantic. Geophysical Research Letters, 29 (17), 1835.

Anderson, L.A., 1995. The hydrogen and oxygen content of marine phytoplankton. Deep-Sea Research 42, pp. 1675-1680.

Anderson, L.A., Sarmiento, J.L., 1994. Redfield ratios of remineralization determined by nutrient data analysis. Global Biogeochemical Cycles 8, pp. 65-80.

Antia, A.N., Koeve, W., Fischer, G., Blanz, T., Schulz-Bull, D., Scholten, J., Neuer, S., Kremling, K., Kuss, J., Peinert, R., Hebbeln, D., Bathmann, U., Conte, M., Fehner, U., Zeitschel, B., 2001. Basin-wide particulate carbon flux in the Atlantic Ocean: Regional export patterns and potential for atmospheric CO_2 sequestration. Global Biogeochemical Cycles, Vol.15, No. 4, pp. 845-862.

Cooper, D.J., Watson, A.J., Ling, R.D., 1998. Variation of pCO_2 along a North Atlantic shipping route (U.K. to the Caribbean): A year of automated observations. Marine Chemistry 60, pp. 147-164.

Dandonneau, Y., 1995. Sea-surface partial pressure of carbon dioxide in the eastern equatorial Pacific (August 1991 to October 1992): A multivariate analysis of physical and biological factors. Deep-Sea Research II, Vol., 42, pp. 349-364.

Dugdale, R.C., Goering, J.J. (1967). Uptake of new and regenerated forms of nitrogen in primary productivity. Limnology and Oceanography, 12, 196-206.

Friis, K., Körtzinger, A., Wallace, D.W.R. Spectrophotometric pH-measurement in the ocean: Requirements, design and testing of an autonomous charge-coupled device detector System. Limnology and Oceanography, submitted.

Gruber, N., 1998. Anthropogenic CO_2 in the Atlantic Ocean. Global Biogeochemical Cycles, Vol. 12, pp. 165-191.

Houghton, J.T., Ding, Y., Griggs, D.J., Noguer, M., van der Linden, P.J., Dai, X., Mashell, K., Johnson, C.A., 2001, in: Climate Change 2001. The Scientific Basis. Editor: Houghton, J.T., Cambridge University Press, ISBN 0521014956, pp. 185-237.

Johnson, K. M., Körtzinger, A., Mintrop, L., Duinker, J.C., Wallace, D.W.R., 1999. Coulometric total carbon dioxide analysis for marine studies: measurement and internal consistency of underway TCO_2 concentrations. Marine Chemistry, 67, pp. 123-144.

Körtzinger, A., Koeve, W., Kähler, P., Mintrop, L., 2001. C:N ratios in the mixed layer during productive season in the northeast Atlantic Ocean. Deep-Sea Research 48, pp. 661-688.

Körtzinger, A., Koeve, W., Kähler, P., Mintrop, L., 2001. C:N ratios in the mixed layer during productive season in the northeast Atlantic Ocean. Deep-Sea Research I, 48, pp. 661-688.

Körtzinger, A., Mintrop, L., Wallace, D.W.R., Johnson, K.M., Neill, C., Tilbrook, B., Towler, P., Inoue, H., Ishii, M., Shaffer, G., Torres, R.F., Ohtaki, E., Yamashita, E., Poisson, A., Brunet, C., Schauer, B., Goyet, C., Eischeid, G., 2000. The international at-sea intercomparison of fCO_2 systems during the R/V Meteor cruise 36/1 in the North Atlantic Ocean. Marine Chemistry, 72, pp. 171-192.

McGillis, W.R., Edson, J.B., Hare, J.E., Fairall, C.W., 2001. Direct covariance air-sea CO_2 fluxes. Journal of Goephysical Research, vol. 106, NoC8, pp. 16,729 - 16,745.

Mehrbach, C., Culberson, C.H., Hawley, J.E., Pytkowicz, R.M., 1973. Measurement of the apparent dissociation constants of carbonic acid in seawater at atmospheric pressure. Limnology and Oceanography, Vol. 18 (6), pp. 897-907.

Mintrop, L., Pérez, F.F., Davila, M.G., Casiano, J.M.S., Körtzinger, A., 2000. Alkalinity determination by potentiometry – intercalibration using three different methods. Ciencias Marinas, 26, pp. 23-37.

Murphy, P., Nojiri, Y., Fujinuma, Y., Wong, C.S., Zeng, J., Kimoto, T., Kimoto, H., 2001. Measurements of Surface Seawater fCO_2 from Volunteer Commercial Ships: Techniques and Experiences from Skaugran. Journal of Atmospheric and Oceanic Technology, Vol. 18 (10), pp. 1719-1734.

Nightingale, P.D., Malin, G., Law, C.S., Watson, A.J., Liss, P.S., Liddicoat, M.I., Boutin, J., Upstill-Goddard, R.C., 2000. In situ evaluation of air-sea gas exchange parameterizations using novel conservative and volatile tracers. Global Biogeochemical cycles, Vol. 14, NO. 1, pp. 373-387.

Omar, A., Johanessen, T., Kaltin, S., Olsen, A. the anthropogenic increase of oceanic $pCO2$ in the Barents Sea Surface water. In press, Journal of Geophysical Research, Oceans.

Pilson, M.E.Q., 1998. An introduction to the chemistry of the sea. Prentice-Hall, Inc.

Redfield, A.C., Ketchum, B.H., Richards, F.A., 1963. The influence of organisms on the composition of sea water. In: Hill, M.N. (Ed.), The Sea, Vol.2. Interscience, New York, pp. 26-77.

Riebesell, U., Zondervan, I., Rost, B., Tortell, P.D., Zeebe, R.E., Morel, F.M.M., 2000. Reduced calcification of marine plankton in response to increased atmospheric CO_2. Nature, Vol.407, pp. 364-367.

Sy, A., 1993, in: Schifffahrt und Meer. Ehlers, P. (Ed.), E.S. Mittler & Sohn GmBH (publisher), Herford. ISBN: 3813204030, pp. 229-237.

Wallace, D.W.R., 2001. Storage and Transport of excess CO_2 in the Oceans: The JGOFS/WOCE Global CO_2 Survey. In: Ocean Circulation and Climate. ISBN 0-12-641351-7, pp. 489-521.

Wanninkhof, R., 1992. Relationship between gas exchange and wind speed over the ocean. Journal of Geophysical Research, 97, pp. 7373-7381.

Wanninkhof, R., McGillis, W.R., 1999. A cubic relationhip between air-sea CO_2 exchange and wind speed. Geophysical Research Letters, Vol. 26, NO 13, pp. 1889-1892.

Wong, C.S., chan, Y.-H., 1991. Temporal variations in the partial pressure and flux of CO_2 at Ocean Station P and Alert, Canada. Tellus B, 43, 206-223.

Zeebe, R.E., Wolf-Gladrow, D., 2001. CO_2 in Seawater: Equilibrium, kinetics, isotopes. Elsevier Science B.V. ISBN: 0444509461 (paperback).

The pCO$_2$ Variability in the North Atlantic Ocean

Heike Lüger*

(corresponding author)

Douglas W.R. Wallace*

Arne Körtzinger*

Yukihiro Nojiri[§]

*Institut für Meereskunde an der Christian-Albrecht-Universität zu Kiel, Düsternbrooker Weg 20, 24105 Kiel, Germany

[§]National Institute for Environmental Studies, 16-2, Onogawa, Tsukuba, Ibaraki, 305-0053, Japan

Abstract

The results of one year of automated pCO$_2$ measurements onboard the car carrier M/V Falstaff are presented and analyzed with regard to the driving forces that change the seawater pCO$_2$ in the mid-latitude North Atlantic ocean. Our dataset shows that the western basin displays a higher variability of seawater pCO$_2$ than the eastern basin. The seasonal drawdown of nitrate is higher in the eastern basin than in the western basin. In both basins, however the seasonal cycle of nitrate is in opposite phase to the temperature cycle. On the other side it is obvious that mixed layer depth and nitrate are correlated as they show similar seasonal cycles in the eastern and western basin, respectively. Our results underline furthermore that we cannot estimate the biomass and subsequently the pCO$_2$ from chlorophyll a. In the eastern basin the temperature and biology effect cancel each other and the overall effect is the observed subdued seasonal cycle of the seawater pCO$_2$. In the western basin the biological drawdown is gradual and it does not counteract the temperature effect. The western basin is a regime where the pCO$_2$ is more temperature controlled than in the eastern basin.

The variability of surface pCO₂ in the North Atlantic Ocean by Heike Lueger

1. Introduction

Since the onset of major agricultural changes and the industrial revolution the CO_2 content in the atmosphere has increased by 30%. CO_2 is the most important anthropogenic greenhouse gas and leads to increased radiative forcing [*Houghton et al.*, 1996], thus potentially changing the global climate. Possible impacts of this global climate change cannot be fully appreciated as of now; one of the reasons is the limited understanding of the global carbon cycle and the role of the ocean. A central question is how much anthropogenic CO_2 ocean and land take up now and in the future. Current estimates of the uncertainty of the ocean uptake term are 0.8 PgC yr^{-1} which is a fourfold uncertainty compared to atmospheric measurements [cf. *Wallace*, 2001]. Calculation of the oceanic CO_2 flux from the atmosphere-ocean pCO_2 is not straightforward and involves many sources of error. Furthermore, the variability of the seawater pCO_2 is difficult to assess, because direct measurements are sparse. It has been the focus of many research projects to explain the variability of oceanic pCO_2 by a number of parameters: sea surface temperature [*Lefèvre and Taylor*, 2002, 1994; *Boutin et al.*, 1999; *Stephens et al.*, 1995], SST anomaly [*Etcheto et al.*, 1999], sea surface salinity and temperature [*Weiss et al.*, 1982], physical transports [*Dandonneau*, 1995; *Winn et al.*, 1994], air-sea exchange [*Lefèvre et al.*, 1994] and phytoplankton blooms [*Watson et al.*, 1991; *Takahashi et al.*, 1993]. Predicting the pCO_2 from changes in SST and SSS is generally most successful in oligotrophic areas such as the subtropical gyres of the North Pacific and Atlantic, especially if the correlation is resolved seasonally [*Bates et al.*, 1996]. In areas with higher biological productivity such as temperate and subpolar seas however, the pCO_2 is more strongly affected by biological drawdown [*Watson et al.*, 1991]. Vertical mixing also plays an important role in these areas as has been shown for the North Pacific [*Sasai et al.*, 2000].

All approaches that are concerned with the variability of the seawater pCO_2 show the need of thorough data sampling. Much effort has been raised to establish an ocean wide network that can monitor changes in pCO_2 and related parameters. Especially in the Pacific, results from regular measurements onboard commercial vessels are promising [*Nojiri et al.*, 1999; *Zeng et al.*, 2002]. In the Atlantic ocean a comparable network was initialised by the European project CAVASSOO [Carbon Variability Studies by Ships of Opportunity] in 2001. This project encompasses three commercial and one research vessel on which seawater and atmospheric pCO_2, SST, and SSS are continuously monitored. All lines are operating with onboard pCO_2 measurement units that collect data continuously with the collected information having greatly increased the present data base. Furthermore, on many cruises discrete samples such as nutrients, chlorophyll *a* etc. have been collected. This includes data for the winter months for which such data are extremely sparse.

This work presents results of one year of measurements onboard one of these commercial vessels: the car carrier M/V Falstaff. The data are used to calculate the seasonal development of the seawater pCO_2 within the eastern and western basin of the North Atlantic. Correlations between the seawater pCO_2 and SST / nutrients and other parameters are examined and used to explain the seasonal variability. Also, basin-wide estimates of the temperature, biology, and air-sea gas exchange effects on the seawater pCO_2 are made in an attempt to explain their variability throughout the year. We compare and discuss our observations with reference to the climatology of *Takahashi et al.* [2002] which is based on historical data.

2.   Methods

2.1.   Data Collection and Analytical Methods

In January, 2002, the M/V Falstaff was outfitted for automated shipboard measurements of CO_2 and closely related parameters. From February, 2002, to January, 2003, the M/V Falstaff sailed across the North Atlantic 15 times. One roundtrip usually took 6 weeks including several stops in Europe and the US with a mean cruising velocity of 18 knots. Continuous measurements of seawater pCO_2, temperature, and salinity were recorded on all cruises. The nautical data (UTC-time, position) were retrieved from a GPS receiver that was installed on the upper deck of the ship. Here we also installed the air inlet from where we pumped the air for the determination of atmospheric CO_2 down to the measurement unit located in the ships engine room. In addition to the continuous measurements we also manually collected discrete samples for nutrients and chlorophyll *a* on every second cruise (Table 1).

The seawater inlet is located at the ship's starboard sea chest located in the engine room of the ship (approximately 7m below the waterline). Temperature and salinity are measured from a thermosalinograph (model 21, Seacat from Seabird Electronics Inc., Seattle/USA) with the remote temperature sensor installed next to the seawater inlet. The SST and SSS data were recorded every 6 seconds and merged with the navigation data. The data are subsequently stored as 1-minute averages.

After passing the thermosalinograph the seawater flows into the pCO_2 measurement unit. This system was designed and loaned to us by Y. Nojiri from the National Institute of

The variability of surface pCO$_2$ in the North Atlantic Ocean by Heike Lueger

Environmental Studies and Kimoto Inc. (both Tsukuba, Japan). The seawater flows at a rate of approximately 20 l/min into the equilibrator where thermodynamic equilibrium is reached with a countercurrent air flow. The equilibrator (Japanese Patent No. P2001-83053A) is a tandem type that includes two stages. The first stage is a static mixer equilibrator with a large

Table 1: Falstaff cruises between February, 2002, and January, 2003. On all cruises we measured seawater/atmospheric pCO$_2$, SST, and SSS continuously. Nuts: discrete nutrient samples; Chl: discrete chlorophyll a samples.

Cruise #	time	Maximum Latitude	Minimum Latitude	Minimum Longitude	Maximum Longitude	samples (discrete)
FAL01	Feb, 2002	50°N	41°N	2°W	67°W	Nuts / Chl
FAL02	Mar, 2002	41°N	38°N	40°W	67°W	-
FAL03	Apr, 2002	49°N	41°N	20°W	71°W	Nuts / Chl
FAL04	May, 2002	50°N	37°N	12°W	60°W	-
FAL05	May, 2002	50°N	41°N	5°W	71°W	Nuts / Chl
FAL06	Jun, 2002	44°N	35°N	10°W	73°W	-
FAL07	Jul, 2002	50°N	41°N	5°W	70°W	Nuts / Chl
FAL08	Jul, 2002	43°N	34°N	13°W	73°W	-
FAL09	Aug, 2002	50°N	41°N	5°W	73°W	Nuts / Chl
FAL11	Sep, 2002	50°N	41°N	4°W	73°W	Nuts / Chl
FAL12	Oct, 2002	44°N	33°N	5°W	75°W	-
FAL13	Nov, 2002	51°N	40°N	1°W	74°W	Nuts / Chl
FAL14	Nov, 2002	44°N	33°N	29°W	80°W	-
FAL15	Dec, 2002	50°N	41°N	5°W	70°W	Nuts / Chl
FAL16	Jan, 2002	44°N	35°N	10°W	75°W	Nuts / Chl

water inlet to prevent clogging and the second stage is a bubble type equilibrator. The over-all equilibration efficiency during the air passage is >99.5% (*Nojiri*, personal communication). The pCO$_2$ system is open to the atmosphere, thus preventing any pressure gradient that might bias the pCO$_2$ determination. The seawater finally flows into a waste tank where it is actively pumped to the outside of the ship.

The countercurrent air flow from the air pump is equilibrated with the seawater pCO$_2$ and subsequently pumped into the analysis unit. After equilibration and prior to analysis the gas is dried in several steps that include airfilters, Peltier elements, Nafion® tubing, and a magnesium perchlorate trap. Finally, a non-dispersive infrared (NDIR) detector (LiCOR®, model 6252, Lincoln, USA) measures the mole fraction of CO$_2$ in the gas sample every minute. The analysis cycle involves a calibration cycle every six hours where a suite of three standard gases (250, 350 to 450 ppm CO$_2$ in natural air) is measured. These working standard gases have been calibrated against NOAA primary standards with similar concentration range using a LiCOR® NDIR analyzer (model 6262). The NOAA gases have an accuracy of 0.07 ppm (cited from: World Meteorological Organisation). After each calibration run, atmospheric air is measured for 20 minutes which is repeated every 2 hours. The seawater pCO_2 is measured in the time left (approximately 880 min/day). For data analysis the raw voltage readings of the NDIR are corrected for temperature and pressure effects following the procedures of the DOE-handbook [*DOE*, 1994] using a least squares procedure for the quadratic regression function. On 10 cruises throughout the year nutrient samples were taken every 3 hours and stored in a freezer onboard the ship (Table 1). These samples were analyzed in the shore-based laboratory at the IfM, Kiel, following the method of *Hansen and Koroleff* [1999]. Chlorophyll a samples (1-2 litres) were taken every 6 hours. The water was filtered onboard using glass fiber filters (GF/F) which were also stored in a freezer until analysis at the IfM, Kiel. Chlorophyll *a* was determined at the IfM, Kiel, using the spectrophotometric method by *Jeffrey and Humphrey* [1975].

2.2. Data Analysis

2.2.1. Grid bands

One year of observations is used to create seasonal maps of seawater and atmospheric pCO_2 and SST. The data from each cruise were first averaged into 12 meridional bands of 5° longitudinal widths and varying zonal extent (Figure 1). The dataset had a maximal latitudinal range of 10 degrees. Ocean regions east of 10°W and west of 70°W were excluded from the calculation in order to remove coastal influences.

After *Zeng et al.* [2002] the seasonality of both the seawater and atmospheric pCO_2 and SST were analyzed with a sigmoidal function which can be expressed by a harmonic equation. The latter was computed for each 5 degree longitude band:

$x(t) = c_0 + c_1 \sin(2\pi t) + c_2 \cos(2\pi t) + c_3 \sin(4\pi t) + c_4 \cos(4\pi t)$ (1)

where x is the seasonally varying quantity (e.g. pCO_2), t is time (t = month) and $c_1 - c_4$ are four seasonal terms. It is required that for the seasonal coverage the maximum data gap should not exceed three months.

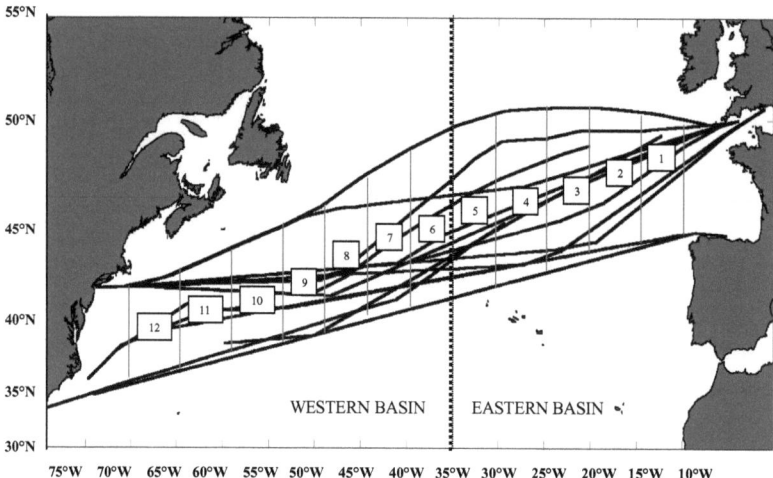

Figure 1: Cruises of M/V Falstaff between February, 2002, and January, 2003. The grey lines indicate the limits of 12 grid bands that were used in the data analysis for estimating seasonal cycles of pCO_2 and related parameters from observations. Marginal seas were excluded (see text). The dotted line indicates the boundary between the eastern (10°W-35°W) and western (36°W-70°W) basin.

To estimate the statistical error of this fitting procedure we compared the seasonal function with the observed data (Figure 2a). The regression of the geometric mean of observed and fitted data yields the following equation:

$(pCO_{2\ fitted}) = 0.83\ (\pm 0.04)\ pCO_{2\ observed} + 58.2\ (\pm 13)$ $R^2 = 0.83$ (2)

There is a negligible offset in the residuals (mean = 0.00004) and the standard deviation of the difference is $\pm$ 9.47 (Figure 2b). The distribution of the residuals (fitted – observed data) shows no trend within the fitting method, neither in space or time (not shown). We also calculated the concordance correlation coefficient according to *Lin* [1989] which helps to evaluate the reproducibility of the fitting method by measuring the variation from the 45° line. We retrieved a coefficient of $\rho c = 0.82$ which indicates that the agreement between observed and fitted data is good.

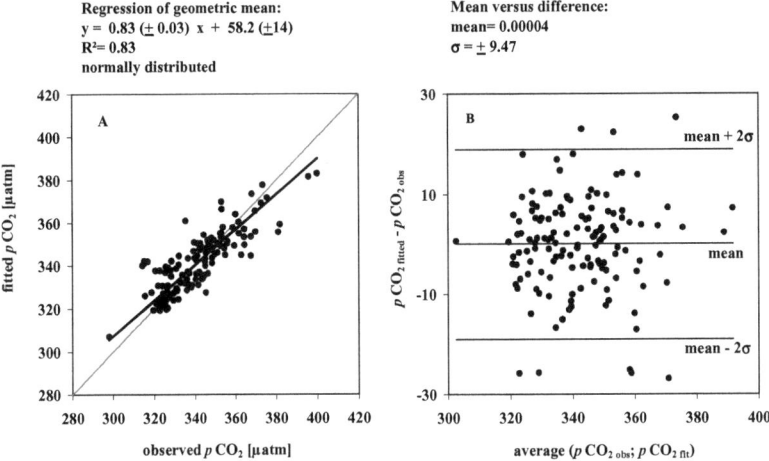

Figure 2: Comparison of observed and fitted seawater pCO_2. The observed data were averaged to 5° longitudinal grids of varying zonal extent from the continuous data. The fitted data were computed using equation 1. A: The geometric mean of observed and fitted data yields: y = 0. 83x +58.2 (R^2 =0.83). Both observed and fitted data follow the normal distribution. B): Residuals plotted against the mean of observed and fitted. The black lines indicate the mean and the 2σ margins.

2.2.2. Temperature and non-temperature effects on the seawater pCO_2

An objective of this work is to determine and quantify temperature and non-temperature related effects on surface seawater pCO_2 in the mid-latitude Atlantic Ocean. We calculate the seasonal amplitudes of the pCO_2 which helps distinguish temperature and non-temperature forced regimes [*Takahashi et al.*, 2002]. As a convention a negative pCO_2 amplitude is assigned if the annual maximum pCO_2 is found during winter (factor: -1). A positive pCO_2 amplitude is assigned if the annual maximum pCO_2 value is found during summer (factor: +1). Note that winter and summer are defined as the 6 month periods from November to April and May to October, respectively.

The results of this computation using the fitted data is displayed in Figure 3. In the eastern North Atlantic the seasonal pCO_2 amplitude ranges between –21 and –30 µatm whereas in the western North Atlantic the results are positive (+28 to +74 µatm). Based on this approximation we separate the North Atlantic into two regimes: an eastern and a western basin comprising the region from 10°W to 34.9°W and 35°W to 70°W, respectively.

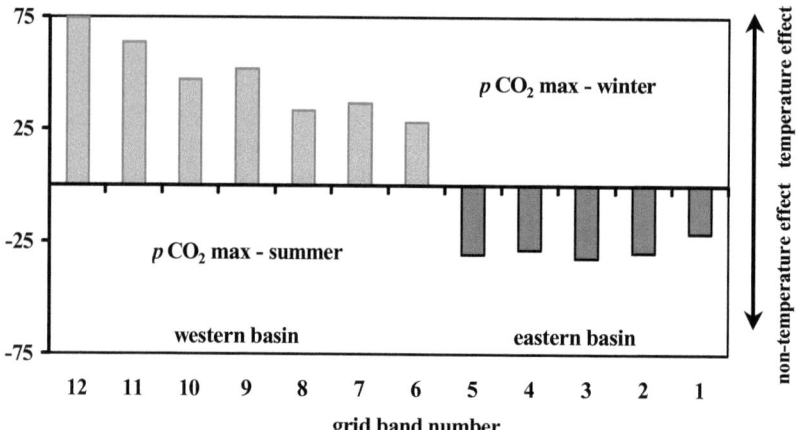

Figure 3: Amplitude of the seawater pCO_2 in the eastern and western basin. The data are computed according to *Takahashi et al.* [2002]. The light grey colour indicates areas where the maximum seawater pCO_2 occurs in the warm season (here: western basin). The dark grey colour refers to areas where the maximum seawater pCO_2 occurs during the cold season (here: eastern basin). The y-axis of the bar plot denotes the seasonal amplitude of seawater pCO_2 and the x-axis labels the averaged grid boxes.

Next we parameterise temperature and non-temperature related effects on the seawater pCO_2 according to the following equations [*Takahashi et al.*, 2002]:

Temperature dependent pCO_2:

$$pCO_2 \text{ at } T_{obs} = pCO_{2\ mean} * \exp[0.0423 * (T_{obs} - T_{mean})] \qquad (3)$$

Temperature independent pCO_2:

$$pCO_2 \text{ at } T_{mean} = pCO_{2\ obs} * \exp[0.0423 * (T_{mean} - T_{obs})] \qquad (4)$$

where $pCO_{2\ mean}$ denotes the observed annual mean seawater pCO_2 per grid band, $pCO_{2\ obs}$ represents the observed monthly mean seawater pCO_2 per grid band, and T_{mean} and T_{obs} are the annual and observed mean SST per grid band, respectively. Equation (3) corrects the annual mean pCO_2 values for effect of temperature on the solubility and the speciation of dissolved CO_2 (4.23% change per 1° Celsius; [*Takahashi et al.*, 1993]) and shows the changes in pCO_2 that are driven by temperature only. Equation (4) corrects the observed seawater pCO_2 for the temperature effect and thus reveals the non-temperature related effects. This leads to seawater pCO_2 values that depend on a range of properties such biology, air-sea gas exchange, or mixing events.

2.2.3. Temperature, biology, and air-sea exchange effects on the seawater pCO_2

In this section we estimate the effects of temperature change, biological drawdown, and air-sea exchange of CO_2 on the ΔpCO_2 (= pCO_2 seawater - pCO_2 atmosphere) in the eastern and western basins of the North Atlantic following the basic approach of *Lefèvre et al.* [1994]. The thermodynamic effect of temperature on the pCO_2 is calculated using equation 3. The ΔpCO_2 (T) for a specific month was computed from the difference between the result of (3) and the observed pCO_2 from the previous month.

The biological drawdown and its effect on the seawater pCO_2 is calculated from the nitrate change per month using a Redfield ratio of 7.2 for the North East Atlantic [*Körtzinger et al.*, 2001]:

$$\Delta pCO_2 \text{ (bio)} = 7.2 \ast (\Delta NO_3 / DIC_{mean}) \ast R \ast pCO_{2\,pm} \tag{5}$$

where ΔNO_3 is the monthly change in nitrate concentration, DIC_{mean} is the annual mean value of dissolved inorganic carbon for each basin, R is the buffer factor calculated for each basin, and pCO_{2pm} denotes the observed seawater pCO_2 of the previous month. Both DIC_{mean} and the buffer factor were calculated from the pCO_2, total alkalinity (TA), SST and SSS annual mean data for each basin. The buffer factor was slightly higher in the eastern (10.56) than in the western basin (10.02). The alkalinity data in turn are derived from the salinity observations (SSS_{obs}) based on a linear regression of discrete TA (µmol/kg) and SSS data collected over the sampling period:

$$TA\ [\mu mol/kg] = 50.78 \ast SSS_{obs} + 527 \tag{6}$$

This regression compares very well with earlier observations [*Körtzinger et al.*, 2001; *Millero et al.*, 1998]. The resulting ΔpCO_2 (bio) represents the change in pCO_2 based on changes in nitrate concentrations and describe the evolution of the pCO_2 driven only by new production.

In order to calculate the air-sea exchange effect on the observed pCO_2 (ΔpCO_2 (ASE)), the current dataset was combined with climatological data for the mixed layer depth and wind speed. The mixed layer depth (MLD) data are from the Levitus 1994 climatology [*Levitus*, 1994]. Here, the MLD is defined as the depth where the potential density is 1.25×10^{-4} g/cm^3 greater than the surface potential density. The wind speed data were monthly mean averages from the COADS climatology. We use wind speeds and a transfer coefficient that is based on the long-term parameterization of *Wanninkhof* [1992] to calculate the CO_2 flux. Subsequently the ΔpCO_2 associated with gas exchange is calculated:

$$\Delta pCO_2 \text{ (ASE)} = R \ast pCO_{2\,pm} \ast F / (DIC_{mean} \ast MLD) \tag{7}$$

where R is the annual buffer factor for each basin, $pCO_{2\ pm}$ is the observed seawater pCO_2 of the previous month, F is the CO_2 flux, DIC is the annual mean value of the dissolved inorganic carbon for each basin, and MLD is the mixed layer depth.

3. Results

3.1. Seasonal cycles of the surface seawater pCO_2, temperature, and nutrients

We present monthly pCO_2 maps in the mid-latitude North Atlantic from February, 2002 to January, 2003 (Figure 4). From January to March the seawater pCO_2 ranges between 320 and 340 µatm with slightly lower values in the western basin than in the eastern basin. In April the eastern basin shows higher values (345-360 µatm) than the western basin (300-330 µatm) indicating an earlier onset of the productive season in the west. In May and June the seawater pCO_2 starts to decrease in the eastern basin (340 to 330 µatm), whereas in the western basin the pCO_2 increases slightly (325 to 335 µatm). From July to September the pCO_2 values rise significantly in the western basin which is attributed to warming. During the summer months the difference of pCO_2 between the eastern and western basins is about 30 µatm. In October and November the pCO_2 of both basins is distributed more uniformly with values around 340 µatm. In December the pCO_2 values are higher than in the previous two months which is likely the result of vertical mixing entraining subsurface waters with higher pCO_2. The western basin shows a higher variability of seawater pCO_2 than the eastern basin (Figure 5 a, b). Generally, the seawater pCO_2 is lower than the atmospheric pCO_2 (370.3 ± 5 µatm) except during the summer months in the western basin. Therefore this part of the ocean (34°N–50°N) mostly represents a sink for atmospheric CO_2. The SST data show similar patterns of variability (Figure 5 c, d) for each basin, with higher seasonal amplitude in the western basin (14 °C) than in the eastern basin (7 °C) when calculated from the observed monthly means. The nitrate data are shown for the eastern basin and the western basin, respectively (Figure 5e, f). The seasonal drawdown of nitrate (winter-summer difference) is higher in the eastern basin (7 µmol/kg) than in the western basin (5 µmol/kg). In both basins, however the seasonal cycle of nitrate is in opposite phase to the temperature cycle. On the other side it is obvious that MLD and nitrate are correlated as they show similar seasonal cycles in both basins (Figure 5 g, h). The eastern basin displays a much broader MLD range than the western basin reaching maximum depth of 360 m in March and April. In the western basin the MLD is generally shallower with maximum depth in March of about 140 m.

To understand the pCO_2 seasonality within each basin the biogeochemical and physical forcing is examined followed by a basin-wide calculation of the biological and air-sea exchange effects on the seawater pCO_2. First we explore the relationship between pCO_2 and SST in the two basins for all cruises (Figure 6). Note, that in both basins no correlation with temperature can be found if all data are considered ($r^2 < 0.12$). We also show the isochemical relationship between pCO_2 and SST based on an empirical relationship 4.23 % /°C that was calculated for North Atlantic seawater samples [*Takahashi et al.*, 1993]. In the western basin

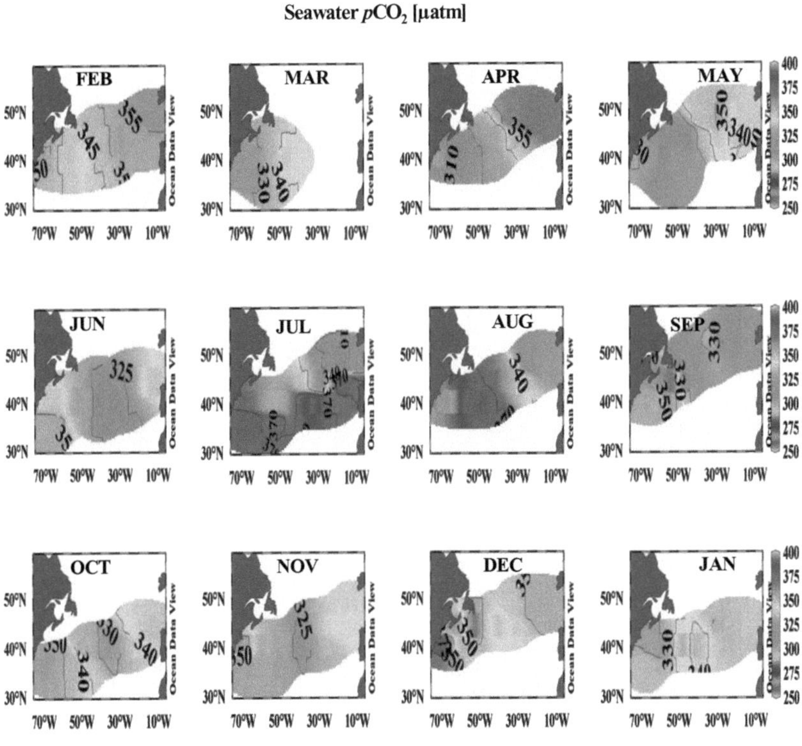

Seawater pCO_2 [µatm]

Figure 4: Monthly distribution of the seawater pCO_2 from February, 2002, to January, 2003. No data were recorded for the eastern basin in March 2002 due to technical problems.

there are cold water masses that yield a negative correlation between pCO_2 and SST (r^2 = 0.32). These data are attributed to Labrador current waters having low SSTs (< 13°C). The correlation between SST and pCO_2 is positive during the summer months (r^2 = 0.53). A positive correlation in the western basin between SST and pCO_2 has been observed previously in the central subtropical gyre of the North Atlantic where the pCO_2 closely follows the mentioned 4.23 %/°C isochemical effect of temperature [*Bates et al.*, 1998]. Our data show a relationship of approximately 1.7 %/°C, and is therefore significantly lower.

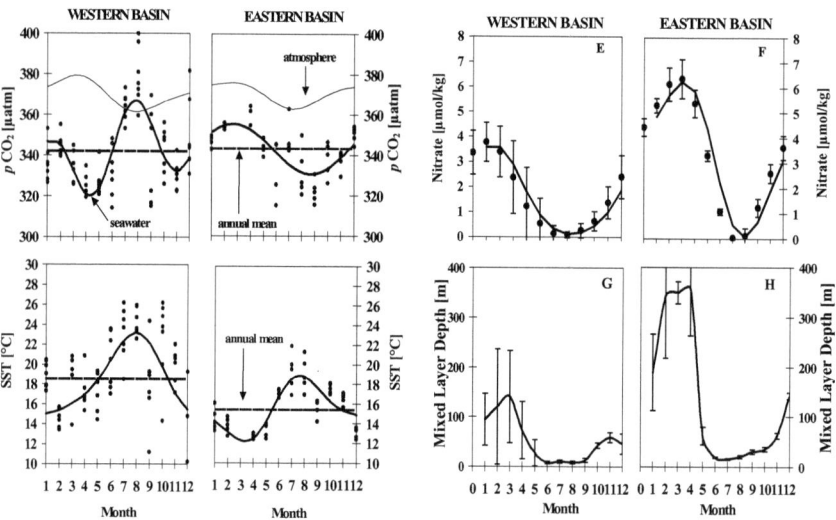

Figure 5: The seasonal cycles of the seawater (a) and atmospheric pCO_2 (b), SST (c), (d), nitrate (e), (f) and MLD (g), (h) for the eastern and the western basin. Shown are the observed data (black dots) and the results of the harmonic computation of each 5° longitudinal grid in the basins.

The variability of surface pCO_2 in the North Atlantic Ocean by Heike Lueger

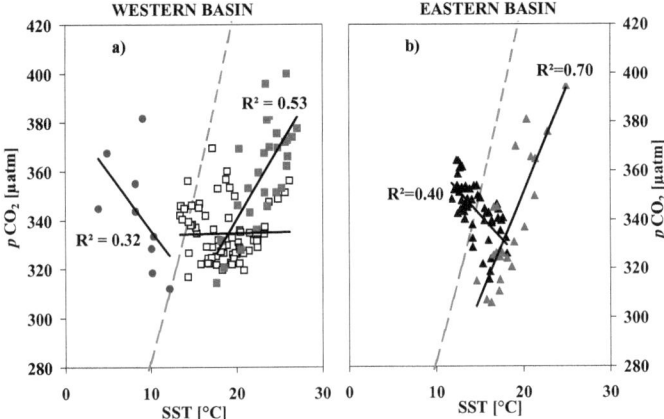

Figure 6: Property-to-property plot of SST and seawater pCO_2 for (a) the western and (b) the eastern basin. In the western basin we found no correlation between SST and pCO_2 (white squares, $R^2 = 0.0004$) except for the following data subsets. We discriminate data with SST lower than 13°C, presumably belonging to Labrador current (blue diamonds) yielding: y = - 4.74 x + 383.4, $R^2 = 0.32$, and data observed in the summer months (red squares): y = 5.94 x + 221.59, $R^2 = 0.53$. This relationships corresponds to 1.7% /°C. In the eastern basin data observed during the summer months (red triangles) resulted in the following regression equation: y = 8.71 x + 176.83, $R^2 = 0.70$ (2.51 %/°C). The black triangles represent the rest of the data set in the eastern basin which yielded: y = - 3.97 x + 401.9, $R^2 = 0.40$. Also shown is the isochemical line (dashed) which describe the empirical relationship of 4.23 %/ °C between pCO_2 and SST that was determined for a North Atlantic seawater sample [*Takahashi et al.*, 1993].

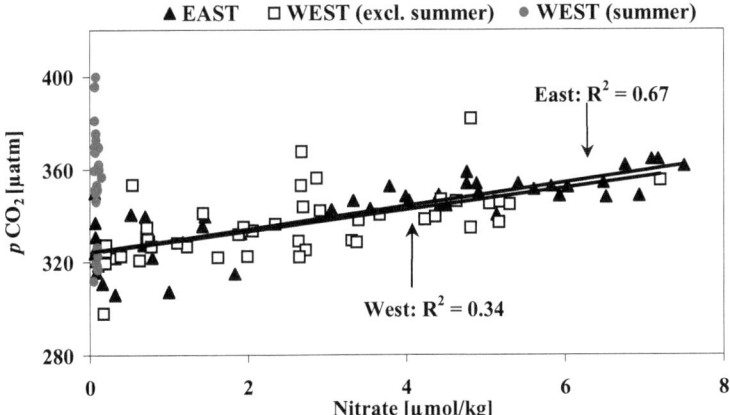

Figure 7: Property-to-property plot of nitrate and seawater pCO_2 for the eastern and the western basin. Black triangles indicate data from the eastern basin and squares (red and white) indicate data from the western basin. In the western basin data from June until September with a nutrient concentration lower than 0.12 μmol/kg are considered separately (red squares) as they yield no correlation with the pCO_2 data (summer depletion of nutrients).

In the eastern basin we also detect a thermodynamic effect on the seawater pCO_2 during summer which is closer to the isochemical effect (Figure 6b). These values yield a regression coefficient of $r^2= 0.70$ and a positive slope of 9.5 which corresponds to an increase of 2.5 %/ °C. A second subset of data in the eastern basin excludes the summer months and shows a negative correlation between SST and pCO_2 ($r^2 = 0.40$). A negative correlation between SST and pCO_2 has for example been attributed to opposing effects such as bloom situations [*Watson et al.*, 1991]. However, a negative relationship between SST and pCO_2 is also found when the mixed layer depth deepens, SST decreases and pCO_2 increases due to entrainment of subsurface water. The high pCO_2 values of these water masses result from their respiratory CO_2 content.

The relationship between seawater pCO_2 and nitrate is analyzed in Figure 7. In the eastern basin the regression coefficient is higher ($r^2 = 0.67$) than in the western basin. Additionally, in the eastern basin the correlation between pCO_2 and nitrate concentration is positive in contrast to the pCO_2-SST relationship. This shows that in the eastern basin SST and nitrate nearly cancel each other with respect to their effects on the seawater pCO_2, thus producing the damped seasonal cycle we noted earlier. A closer look at the nitrate data of the western basin reveals that they are not uniformly distributed, but we can discriminate between two subsets. One subset is characterized by low nitrate concentration and varying pCO_2 (red squares). These data correspond to the summer months (June – September) when nutrients are typically depleted. In the other subset (excluding the summer months) we find a positive correlation between pCO_2 and nitrate ($r^2= 0.34$). This pattern looks similar to the eastern basin, although with a weaker correlation.

We also explored the correlation between seawater pCO_2 and chlorophyll *a* concentration (Figure 8). Chlorophyll *a* has been considered to be an indicator for algal biomass and is sometimes observed to anticorrelate with pCO_2, e.g. *Watson et al.* [1991]. The regression result, however, does not show any significant correlation between pCO_2 and chlorophyll *a* for any region or time ($r^2 < 0.2$), not even for the spring time where one would expect a closer relationship. It is most likely that this relationship is 1) very patchlike and 2) heavily dependent on grazing pressure. Clearly in our result we cannot estimate the seawater pCO_2 from chlorophyll *a*.

Generally, the correlation analysis of temperature and nutrients gives insight into how the pCO_2 is influenced by different forcings. In the next section we will test some of the relationships in a more quantitative way.

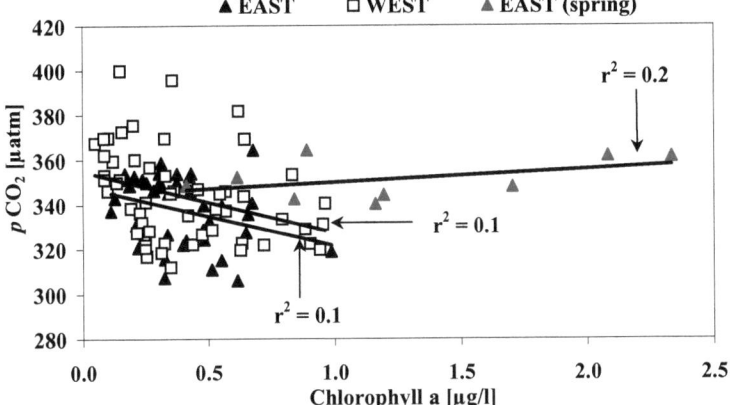

FIG.8: Property-to-property plot of chlorophyll *a* and seawater pCO_2 for the eastern and the western basin. Triangles indicate data from the eastern basin (black and green) and squares indicate data from the western basin. In the eastern basin spring data (April and May) are considered separately (green triangles). We also tested correlation between temperature corrected pCO_2 and chlorophyll a (same dataset), but yielded no significant correlation.

3.2. Seasonal changes in surface seawater pCO_2

Takahashi et al. [2002] created monthly maps of seawater pCO_2 based on historical data collected over 45 years and their results can be compared to this new dataset. In addition we quantify temperature, biology and air-sea exchange effects on seawater pCO_2. The calculation procedures are explained in the methods section.

The difference (T-B) between the thermal (T) and non-thermal (B) forcing were calculated where the latter is considered the biological forcing according to *Takahashi et al.* [2002].

Figure 9 shows the balance between the two. Generally, the eastern basin pCO_2 is dominated by non-temperature effects whereas the western basin pCO_2 is strongly driven by temperature effects. In the eastern basin the pCO_2 values are generally lower during summer than during winter when strong vertical mixing brings up subsurface watermasses with high pCO_2. This can be verified by the mixed layer climatology showing much higher mixed layer depth in winter in the eastern basin than in the western basin as we noted earlier. In the western basin

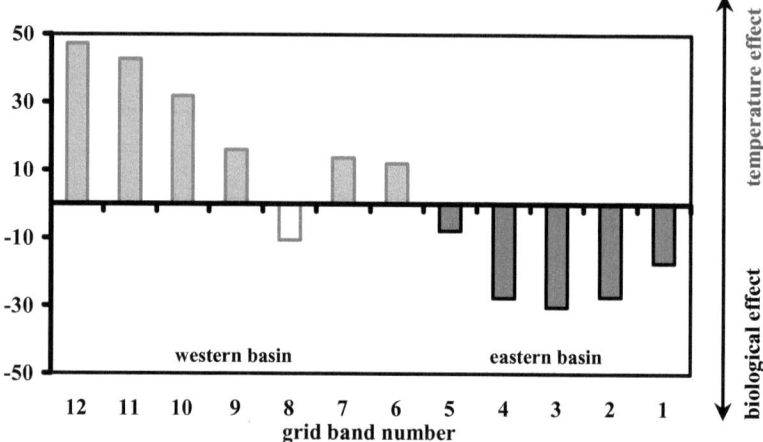

Figure 9: Difference of temperature and non-temperature related effects on the seawater pCO_2 in the eastern and western basin. The data are computed according to *Takahashi et al.* [2002]. The y-axis of the bar plot denotes the relative the temperature and non-temperature effect on seawater pCO_2 and the x-Axis labels the averaged grid boxes.

the pCO_2 is highest in the summer since temperature is the major force that drives the seawater pCO_2 there. What is the biological or non-temperature effect according to *Takahashi et al.* [2002]? By definition it consists of all effects that are not directly related to temperature. It includes biological production and respiration, changes in alkalinity, but also physical processes like vertical mixing and advection as well as the air-sea exchange of CO_2.

We calculate two of these non-temperature effects using the current dataset: the new production (ΔpCO_2 bio) and the air-sea exchange (ΔpCO_2 ASE). The new production is calculated from the observed nitrate concentration. The air-sea exchange term is derived from the CO_2 flux between the ocean and the atmosphere, which means it does not only depend on the ΔpCO_2 but also on MLD and wind speed. For calculation procedures please refer to section 2.2.3.

The seasonal cycle of ΔpCO_2 in the eastern basin seems damped compared to the western basin (Figure 10). The reason for this can be interpreted in terms of the different forcing that influences ΔpCO_2 in each basin. In the eastern basin the subdued ΔpCO_2 cycle is produced by two counteracting effects: temperature and biology (Figure 11 a, c). Here, the $\Delta p CO_2$ (T) with a seasonal amplitude of 57 µatm mirrors the $\Delta p CO_2$(B) which shows a amplitude of 44 µatm.

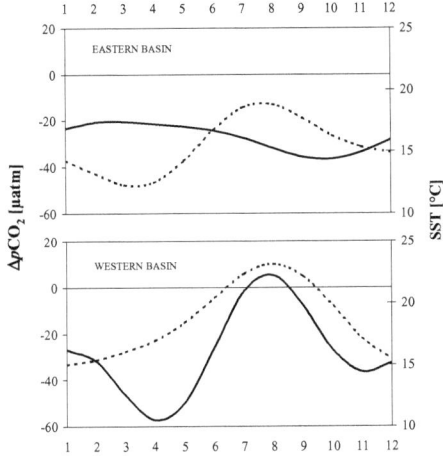

Fig. 10: Annual cycles of observed $\Delta p CO_2$ (black line) and SST (grey dashed line) in the eastern and western basin.

The ASE effect in the eastern basin is only about 50% of the temperature and biological effect (Figure 11 e). It is strongest during the summer time when the mixed layer depth is shoaling thus concentrating the effect of the CO_2 flux. In the western basin the strong seasonal $\Delta p CO_2$ amplitude of 64 µatm is caused by the smaller biological forcing on the $\Delta p CO_2$. In both basins the temperature and ASE forcings are nearly the same and they result in similar seasonal amplitudes, however the timing is different. In the western basin the biological effect is only about 50% of that in the eastern basin (Figure 11 c, d), which is associated with the lower amplitude of nitrate changes and forced by the MLD variability. In the western basin the $\Delta p CO_2$ (ASE) is of the same magnitude as the biology effect albeit of different signs thus both effects are nearly cancelled out. The pronounced maximum of the $\Delta p CO_2$ (ASE) in June is caused by an extremely shallow mixed layer depth (8 m), which might be an artefact of the mixed layer depth climatology.

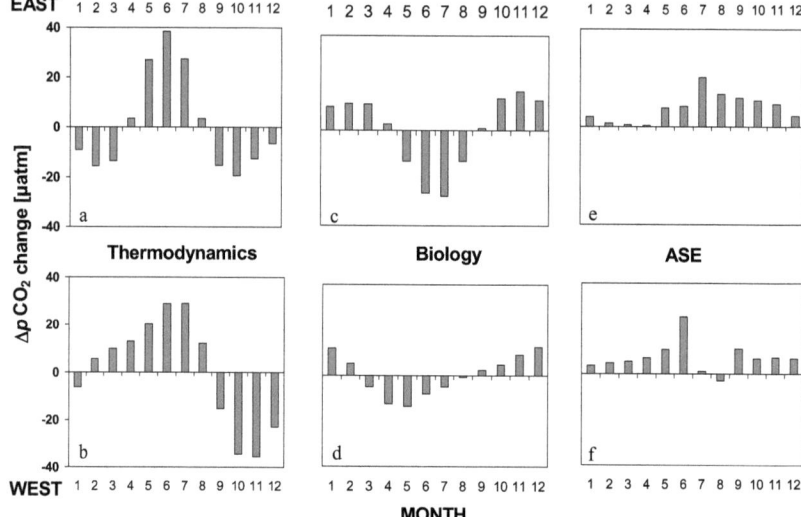

Figure 11: Histograms of the $\Delta p CO_2$ annual cycle in the eastern and western basin calculated from the temperature (thermodynamics; a, b), biology (c,d), and ASE (air-sea exchange; e,f) effect. For calculation procedure please refer to the methods chapter.

We conclude that the seasonal cycle of the seawater pCO_2 is controlled by two major forcings: temperature and biology. In the eastern basin the temperature and biology effect cancel each other and the overall effect is the observed subdued seasonal cycle of the seawater pCO_2. In the western basin the biological drawdown is gradual and it does not counteract the temperature effect. The western basin is a regime that is more temperature controlled than the eastern basin.

4. Discussion

4.1. pCO_2, SST, and nutrient variability in the eastern and western basin of the North Atlantic

Our dataset compares well with earlier observations in this area. *Cooper et al.* [1998] reported one year of observations from a commercial vessel that was equipped with a pCO_2 measurement unit cruising from the U.K. to Jamaica between 1994 and 1995. They found the largest seasonal change of pCO_2 in the mid Atlantic region (40-50°W) with summer supersaturation of pCO_2 which is coherent with our results. Both datasets show that in the North East Atlantic the seawater pCO_2 is undersaturated with respect to the atmosphere. *Cooper et al.* [1998] observe a 2-3 % per °C increase in pCO_2 and explain that this deviation from the thermodynamic effect (4.23 % /°C) is caused by the biological CO_2 uptake. Our observations in the western and eastern basin range between 1.7 to 2.5 % / °C increase in pCO_2 during the summer months, but are significantly lower for the rest of the data set or even in the opposite direction which also can be explained by counteracting biological and ASE effects that decrease the pCO_2. The reasonable pCO_2-SST correlation we observed during the summer months compares to earlier observations in the subtropical gyre of the North Atlantic, especially on a seasonal basis [*Lefèvre et al.*, 2002; *Bates et al.*, 1998]. However, our observations do not support this finding for seasons other than summer. This indicates that the observed water masses along our measured line are 1) different from the subtropical gyre including mixtures of surface water masses different origins and 2) they are more strongly influenced by biological activity, air-sea exchange, and vertical mixing than in the subtropical gyre. Clearly our results show that for this transect it is not possible to predict pCO_2 data from SST changes alone.

The winter-spring correlation between seawater pCO_2 and nitrate for the eastern and western basin reflects the biological forcing. In the eastern basin the biological forcing is stronger based on the correlation between pCO_2 and nitrate. Similar behaviour was also shown by *Lefèvre et al.* [1994] for the equatorial Pacific where they found a very high pCO_2-nitrate correlation ($R^2 > 0.9$) In the western basin of the North Atlantic the temperature forcing is comparable to the eastern basin. However, as the biological forcing is less than in the eastern basin, the temperature effect is not counteracted and the pCO_2 follows primarily temperature changes.

Earlier work suggested a good correlation between pCO_2 and chlorophyll [*Watson et al.*, 1991] which would be very convenient as chlorophyll can be remotely sensed. Our data, however, do not reveal any relationship between pCO_2 and chlorophyll. We could not find any significant correlation, not even during spring time when one might expect that an increase of biomass, often represented as chlorophyll, is accompanied by a decrease in pCO_2. Therefore it will be necessary to find predictors other than chlorophyll that will estimate the seawater pCO_2 in the North Atlantic. A good approach might be the obvious correlation between nitrate and mixed layer depth which will be the topic of a future paper.

4.2. Comparison of pCO_2 changes to Takahashi's climatology

The annual changes of seawater pCO_2 are compared to the pCO_2 changes computed by *Takahashi et al.* [2002]. Both datasets are corrected to the virtual year 1995 as recommended by Takahashi who assumed that ocean areas south of 45°N should be corrected for the annual increase of anthropogenic CO_2 (C_{ant}). Here, we corrected our observed pCO_2 data by using a correction factor of 1.5 ppm/ year for data south of 45°N whereas data north of 45°N were left uncorrected for the anthropogenic increase (C_{ant}). Also shown, however, is the result when all data are corrected for the anthropogenic increase. In eastern basin the climatological data are on average lower by 13 µatm and in the western basin Takashashi's data show a good agreement with the observed data with a mean overestimate of 2 µatm (Figure 12). It is noteworthy that in the eastern basin the climatological estimate is improved once we correct all data for C_{ant}. The differences between the climatology and our observed data are also shown using a latitudinal resolution (Figure 13). There is a latitudinal dependent trend within the comparison. At lower latitudes the climatology overestimates the seawater pCO_2 and at higher latitudes there is an apparent underestimation by Takashashi's climatology compared to observed pCO_2 data. This deviation can not be attributed to a deviation in SST which we also examined; except possibly for the 40°N latitude where the SST deviation between the two datasets is highest. The differences between the Takahashi's climatology and our dataset are most likely caused by two factors. The first one relates to the fact that our dataset only represents one year of data in contrast to Takahashi et al. who used over 900,000 datapoints for all oceans. Secondly, it is most likely that there is indeed an increase in C_{ant} in surface waters north of 45°N in contrast to the assumption by Takahashi et al. (2002). Many

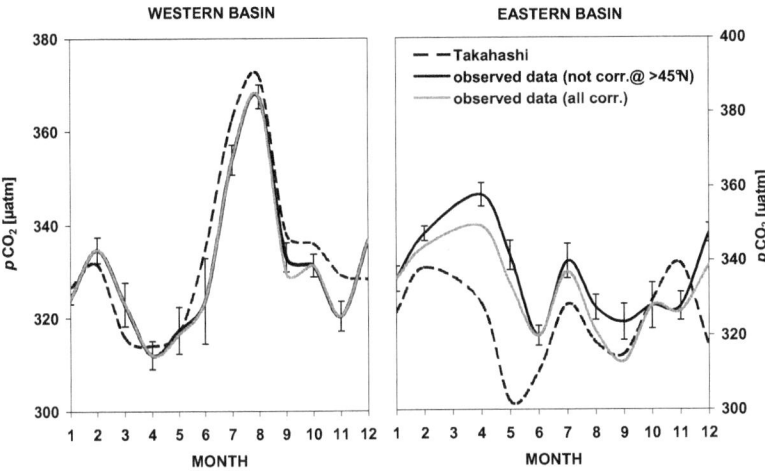

Figure 12: pCO_2 data from Takahashi's climatology [2002] are compared to observed data. The observed data corrected to a virtual year (1995) assuming an annual increase of 1.5 ppm (black line). Also shown are the uncorrected data (grey line).

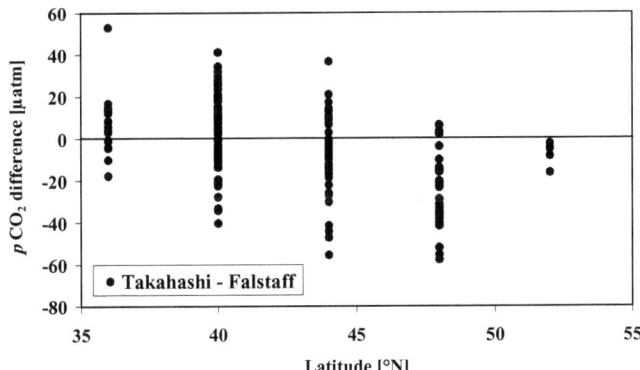

Figure 13: Comparison of observed pCO_2 data and Takahashi's climatology along the latidunal transect. All pCO_2 data were corrected for the annual anthropogenic increase (1.5 ppm / yr) to the virtual year 1995 south of 45°N.

estimates do imply an uptake of C_{ant} in the northern North Atlantic, however, it has not been resolved how strong this uptake is [e.g. *Voelker et al.*, 2002; *Holfort et al.*, 1998].

5. Conclusions

We successfully established a VOS-line in the North Atlantic with the carcarrier M/V Falstaff owned by the Swedish Wallenius Lines A/S. The measurements of seawater pCO_2 and further parameters started in February, 2002, and will continue.

Our results suggest that in the eastern and the western basin the variability of the seawater pCO_2 is steered by different mechanisms. We showed that along our transect predicting the seawater pCO_2 from temperature might only be successful in the summer, but the correlation of pCO_2 and SST is too weak during the rest of the year. Despite earlier observations we showed that chlorophyll cannot be used as a predictor for pCO_2 as we could not prove any correlation between the two. However, our results are promising with regards to the relationship between nitrate and pCO_2. Here the mixed layer depth which is correlated with the nitrate concentration might be an auspicious parameter to be used as a pCO_2 predictor.

It is obvious that working with such a large dataset has advantages with regards to the spatio-temporal resolution, however it also shows that for a complete understanding of the seasonal pCO_2 cycle we need to know the interannual variability. The continuation of scientific measurements onboard commercial vessels is highly recommended as they provide valuable insight biogeochemical workings of the ocean which cannot be achieved by research vessels only.

The variability of surface pCO₂ in the North Atlantic Ocean by Heike Lueger

Acknowledgements

We sincerly thank Per Croner and Sara Gorton of Wallenius Lines, Stockholm/Sweden, for cooperation and generous support. The outstanding support of the chief engineer, Gerth Gulliksson, as well as captain and crew of the M/V Falstaff is greatly appreciated. We also wish to acknowledge Yukihiro Nojiri for generously providing his automated pCO_2 measurement system. We thank Alexander Sy and the German Federal Maritime Agency (BSH) for providing the thermosalinograph. And thank to Peter Fritsche, Kerstin Nachtigall, Hergen Johannsen for chlorophyll and nutrient analysis, Lisa Weber, Philip Nuss, Tobias Steinhoff and Axel Wendt for cruise help, Hans-Peter Hansen for software support, Hela Mehrtens and Thomas Martin for help with the NCEP/NCAR wind speed data. This work was funded by the European Commission under grant no. EVK2-CT-2000-00088.

References

Bates, N. R., A.M. Michaels, A.H. Knap, Seasonal and interannual variability of oceanic carbon dioxide species at the U.S. JGOFS Bermuda Time-series Study (BATS) site, *Deep-Sea Research II, Vol. 43*, 347-383, 1996.

Bates, N. R., T. Takahashi, D. W. Chipman, A. H. Knap, Variability of pCO_2 on diel to seasonal timescales in the Sargasso Sea near Bermuda, *Journal of Geophysical Research, 103*, 15,567-15,858, 1998.

Boutin, J., J. Etcheto, Y. Dandonneau, D. C. E. Bakker, R. A. Feely, H. Y. Inoue, M. Ishii, R. D. Ling, P. D. Nightingale, N. Metzl, Satellite sea surface temperature: a powerful tool for interpreting in situ pCO_2 measurements in the equatorial Pacific Ocean, *Tellus, 51B*, 490-508, 1999.

Cooper, D. J., A. J. Watson, R. D. Ling, Variation of pCO_2 along a North Atlantic shipping route (U.K. to the Caribbean): A year of automated observations, *Marine Chemistry, 60*, 147-164, 1998.

Copin-Montégut, C., A new formula for the effect of temperature on the partial pressure of CO_2 in seawater, *Marine Chemistry, 25*, 29-37, 1988.

Copin-Montégut, C., A new formula for the effect of temperature on the partial pressure of CO_2 in seawater (Corrigendum), *Marine Chemistry, 27*, 143-144, 1989.

Dandonneau, Y., Sea-surface partial pressure of carbon dioxide in the eastern equatorial Pacific (August 1991 to October 1992): A multivariate analysis of physical and biological factors, *Deep-Sea Research II, 42, No. 2-3*, 349-364, 1995.

Dickson, A. G. and C. Goyet (Eds.), *DOE, Handbook of methods for the analysis of the various parameters of the carbon dioxide system in sea water, Report ORNL/CDIAC-74*, Carbon Dioxide Information Analysis Center, Oak Ridge National Laboratory, Oak Ridge, Tennessee, USA. Version 2, 1994.

Etcheto, J., J. Boutin, Y. Dandonneau, D.C.E. Bakker, R.A. Feely, R.D. Ling, P.D. Nightingale, R. Wanninkhof, Air-sea CO_2 flux variabilitiy in the equatorial Pacific Ocean near 100°W, *Tellus, 51B*, 734-747, 1999.

Hansen, H. P., F. Koroleff, Determination of nutrients, in: *Methods of Seawater Analysis*, edited by K. Grasshoff, K. Kremling, M. Ehrhardt, pp. 159-228, Verlag Chemie, Weinheim, 1999.

Holfort, J., K. Johnson, A. Putzka, B.Schneider, G. Siedler, D.W.R. Wallace, The meridional transport of inorganic carbon in the South Atlantic Ocean, *Global Biogeochemical Cycles, Vol. 12*, 479-499, 1998.

Houghton, J. T., L. G. Meira Filho, B. A. Callender, N. Harris, A. K. Kattenberg, K. Maskell, Climate change 1995: The science of climate change, IPCC report, Cambridge University Press, The Edinburgh Building Shaftesbury Road, Cambridge, CB2 2RU, England, 1996.

Jeffrey, S. W., G. F. Humphrey, New spectrophotometric equations for determining chlorophylls a, b, c_1, and c_2 in higher plants, algae and natural phytoplankton, *Biochemie und Physiologie der Pflanzen, 167*, 191-198, 1975.

Keeling, C. D., The concentration and isotopic abundances of carbon dioxide in the atmosphere, *Tellus, 12*, 200-203, 1960.

Körtzinger, A., W. Koeve, P. Kähler, L. Mintrop, C:N ratios in the mixed layer during productive season in the northeast Atlantic Ocean, *Deep-Sea Research I, 48*, 661-688, 2001.

Körtzinger, A., J. I. Hedges, P. D. Quay, Redfield ratios revisited: Removing the biasing effect of anthropogenic CO_2, *Limnology and Oceanography, 46*, 964-970, 2001.

Lefèvre, N., C. Andrié, Y. Dandonneau, pCO_2, chemical properties, and estimated new production in the equatorial Pacific in January-March 1991, *Journal of Geophysical Research, Vol. 99, C6*, 12, 639-12,654, 1994.

Lefèvre, N., A. Taylor, Estimating pCO_2 from sea surface temperatures in the Atlantic gyres. *Deep-Sea Research I, 49*, 539-554, 2002.

Levitus, S., T. Boyer, World Ocean Atlas Volume 4: Temperature, NOAA Atlas NESDIS 4, U.S. Department of Commerce, Washington, D.C., 1994.

Lewis, E., D. W. R. Wallace, Program Developed for CO_2 System Calculations, ORNL/CDIAC-105, Carbon Dioxide Information Analysis Center, Oak Ridge National Labortory, U.S. Department of Energy, Oak Ridge, Tennessee, 1998.

Lin, L. I. K., A Concordance Correlation Coefficient to Evaluate Reproducibility, *Biometrics 45*, 225-268, 1989.

Millero, F. J., K. Lee, M. Roche, Distribution of alkalinity in the surface waters of the major oceans, *Marine Chemistry, 60*, 111-130, 1998.

Nojiri, Y., Y. Fujinuma, J. Zeng, C. S. Wong, Monitoring of pCO_2 with complete seasonal coverage utilizing a cargo ship M/S Skaugran between Japan and Canada/US in: Proceedings of the Second International Symposium of CO_2 in the Oceans, CGER-

Report, CGER-I037-'99, 17-23, National Instiute for Environmental Studies, Tsukuba, Japan, Jan, 1999.

Sasai, Y. M. Ikeda, N. Tanaka, Changes of total CO_2 and pCO_2 in the surface ocean during the mixed layer development in the northern North Pacific, *Journal of Geophysical Research, 105*, No. C2, 3465-3481, 2000.

Siegenthaler, U., J. L. Sarmiento, Atmospheric carbon dioxide and the ocean, *Nature, 365*, 119-125, 1993.

Stephens, M. P., G. Samuels, D.B. Olson, R.A. Fine, Sea-air flux of CO_2 in the North Pacific using shipboard and satellite data. *Journal of Geophysical Research, 100*, No. C7, 13,571-13,583, 1995.

Takahashi, T., J. Olafsson, J. G. Goddard, D. W. Chipman, S. C. Sutherland, Seasonal variation of CO_2 and nutrients in the high.latitude surface oceans: A comparative study, *Global Biogeochemical Cycles, 7*, 843-878, 1993.

Takahashi, T., S. C. Sutherland, C. Sweeney, A. Poisson, N. Metzl, B. Tilbrook, N. Bates, R. Wanninkhof, R. Feely, C. Sabine, J. Olafsson, Y. Nojiri, Global Sea-Air CO_2 flux based on climatological surface ocean pCO_2 and seasonal biological and temperature effects, *Deep Sea Research II, 49 (9-10)*, 1601-1622, 2002.

Voelker, C., D. W. R. Wallace, D. Wolf-Gladrow, On the role of heat fluxes in the uptake of anthropogenic carbon in the North Atlantic, *Global Biogeochemical Cycles, 16*, No. 4, 85-1, 2002.

Wallace, D. W. R., Storage and Transport of Excess CO_2 in the Oceans: The JGOFS/WOCE Global CO_2 Survey, in: *Ocean Circulation and Climate*, 489-521, Academic Press, 2001.

Wanninkhof, R. Relationship between wind speed and gas exchange over the ocean, *Journal of Geophysical Research, 97, (C5)*, 7373-7382, 1992.

Watson, A., C. Robinson, J. Robinson, P.I.B. William, M. Fasham, Spatial variability in the sink for atmospheric carbon dioxide in the North Atlantic, *Nature, 350*, 50-53, 1991.

Weiss, R. F., R. A. Jahnke, C.D. Keeling, Seasonal effects of temperature and salinity on the partial pressure of CO_2 in seawater, *Nature, 300*, 511-513, 1982.

Winn, C. D., F. T. Mackenzie, C. J. Carillo, C. L. Sabine, D. M. Karl, Air-sea carbon dioxide exchange in the North Pacific Subtropical Gyre: Implications for the global carbon budget, *Global Biogeochemical Cycles, 8*, 157-163, 1994.

The variability of surface pCO$_2$ in the North Atlantic Ocean by Heike Lueger

Zeng, J., Y. Nojiri, P. P. Murphy, C. S. Wong, Y. Fujinuma, A comparison of ΔpCO$_2$ distributions in the northern North Pacific using results from a commerical vessel in 1995-1999, *Deep-Sea Research II, 49,* 5303-5315, 2002.

The Seasonality of the Air-Sea CO_2 Flux in the Mid-Latitude North Atlantic

Heike Lüger*
(corresponding author)

Douglas W.R. Wallace*

Arne Körtzinger*

Yukihiro Nojiri[§]

*Institut für Meereskunde an der Christian-Albrecht-Universität zu Kiel, Düsternbrooker Weg 20, 24105 Kiel, Germany

[§]National Institute for Environmental Studies, 16-2, Onogawa, Tsukuba, Ibaraki, 305-0053, Japan

Abstract

We provide a full annual pCO_2 cycle of the latitudinal band (34-50°N) in the Atlantic Ocean. Independent gridding approaches yield a similar result: the North Atlantic ocean was a sink for CO_2 most of 2002. The mean annual CO_2 flux into the ocean for both 1°x1° and 4°x5° gridding methods over the entire area was 6 mmol m^{-2} d^{-1}. We showed that the annual CO_2 flux varied up to 50% depending on which gas transfer equation we used. The 4°x5° dataset with short-term winds yields an annual CO_2 uptake of 0.052 Pg C yr^{-1} which is 30% lower compared to the result using long-term winds (0.071 Pg C yr^{-1}). When comparing the CO_2 uptake rate in the region between 38°N and 46°N between our dataset and the climatology by Takahashi et al. (2002) using the same wind speed and grid resolution, the difference is small (4 %).

The variability of surface pCO$_2$ in the North Atlantic Ocean by Heike Lueger

1. Introduction

The mid-latitude North Atlantic acts is a major sink for atmospheric CO$_2$ most of the year (Lefèvre et al., 1999; Takahashi et al., 2002). This results firstly from the cooling of surface waters and deep water formation in this area; warm salty water masses flow north-eastwards where the water cools and sinks due to increased density (Dickson et al., 1996). Secondly, the carbon dioxide chemistry of the ocean plays an important role in the CO$_2$ uptake. The ocean acts as a buffer towards an increase in atmospheric CO$_2$ because the latter reacts with water to form bicarbonate and carbonate ions in successive reactions. Both the dynamic feature and the carbon chemistry lead to a net uptake and transport of CO$_2$ into the deep ocean. Despite general knowledge of these characteristics uncertainty remains as to the magnitude and temporal variability of CO$_2$ fluxes in the North Atlantic ocean. The uncertainty is a consequence of the variable natural carbon cycle in the ocean which causes seawater pCO$_2$ to vary up to $\pm$ 60% around the atmospheric pCO$_2$ level (Takahashi et al., 2002). Therefore CO$_2$ measurements in the open oceans are in demand in order to analyze and predict the oceanic carbon cycle. Still oceanic estimates of CO$_2$ fluxes are much more constrained than terrestrial CO$_2$ fluxes where the carbon speciation is more diverse which makes it difficult to accurately determine CO$_2$ sources and sinks on land. Therefore it seems reasonable to calculate terrestrial CO$_2$ fluxes via inverse models which are generally based on oceanic and atmospheric observations.

Oceanic research campaigns during winter time are sparse in the North Atlantic ocean, even though international databases have been greatly expanded in recent years. Most databases for the North Atlantic are significantly weighted towards the fair-weather season, which also affects model based estimates and therefore winter-time data are precious and in demand. The EU-funded project CAVASSOO (Carbon Variability Studies by Ship Of Opportunity) started off an Atlantic network of volunteer observing ships (VOS) to monitor seawater pCO$_2$ and related parameters. The VOS line maintained by us is based on the Swedish car carier M/V Falstaff which sails all year round between Europe and the U.S. east coast on a roundtrip of typically six weeks duration. In 2002 the M/V Falstaff sailed the North Atlantic 14 times with an average yield of 6,000 datapoints per cruise and a maximum spatial resolution of 500 m. This database is used here to calculate the regional CO$_2$ flux at two spatial resolutions (4°x 5° and 1°x1°) with different parameterizations of the transfer velocities, and with two sets of wind speed data (NCEP/NCAR and COADS climatology). We also compare our results to the prominent climatology of Takahashi et al. (2002).

The variability of surface pCO₂ in the North Atlantic Ocean by Heike Lueger

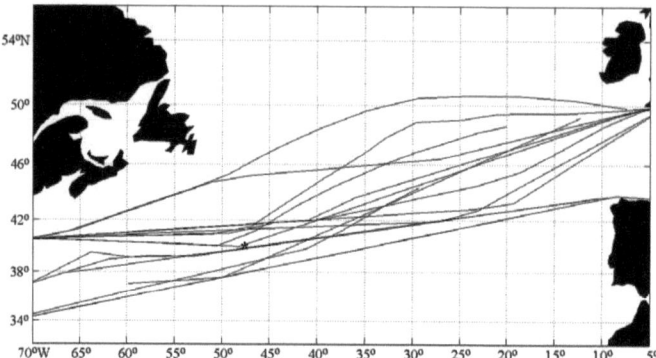

Figure 1: Falstaff cruises for 2002/early 2003 between Europe and the USA. We collected seawater and atmospheric pCO_2, SST, and SSS on all cruises, additionally discrete samples (nutrients, chlorophyll) were collected on every other cruise. The grid boxes indicate the 4°x5° grid used by Takahashi et al. (2002). The asterix in the 40°N, 47.5°W box denotes that this box was used for direct comparison between 4°x5° and 1°x1° gridding.

2. Methods

2.1. Data collection and analytical methods

In early 2002 the carcarrier M/V Falstaff was equipped with an autonomous pCO_2 measurement unit. Here we present data of the first year of almost continuous operation. The M/V Falstaff sailed the North Atlantic 15 times between February, 2002 and January, 2003, collecting about 95,000 pCO_2 datapoints (Figure 1). This corresponds to a spatial resolution of 500 m at an aveage ship speed of 18 knots. In addition to seawater pCO_2 measurements atmospheric pCO_2 was measured every two hours.

The CO_2 system is installed on the starboard side of the lowest deck in the engine room of the Falstaff. The seawater intake is located at a nearby seachest and the seawater flows into the system by hydrostatic pressure only. This avoids bias of the CO_2 measurement that might occur if bubbles form due to cavitation from pump activity. The temperature of the seawater is measured at three locations: (a) *in situ* at the inlet by a remote temperature sensor (model 38, Seacat from Seabird Electronics Inc., Seattle/USA), (b) upstream of the equilibrator by a thermosalinograph (model 21, Seacat from Seabird Electronics Inc., Seattle/USA), and (c) within the equilibrator by a Pt 100 temperature probe.

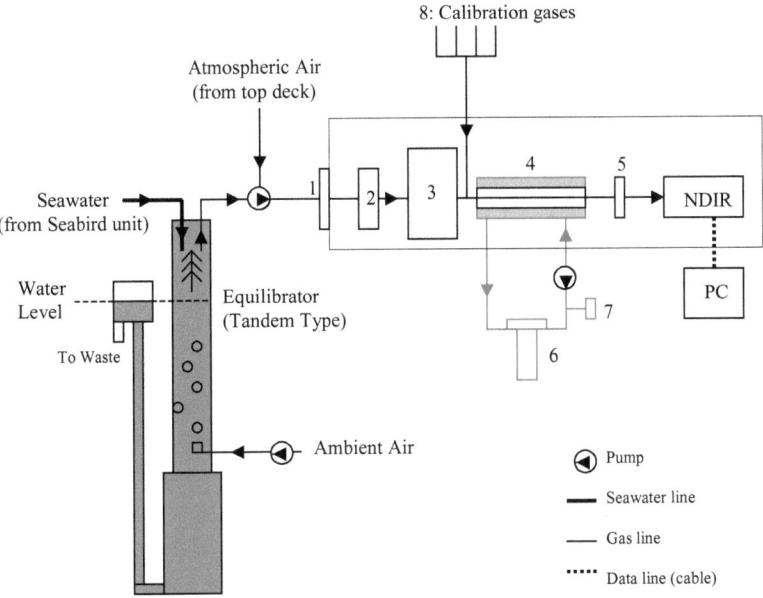

Figure 2: Overview of the pCO$_2$ measurement unit onboard the M/V Falstaff. The instrument was installed in early 2002 in the engine room of the carcarrier M/V Falstaff (appr. 8 m below the waterline). Not shown are the Seabird thermosalinigraph, temperature probe and miscellaneous valves. (1): teflon filter; (2): Mist trap, (3): Peltier cooler, (4): Nafion dryer, (5): Mg(ClO$_4$)$_2$ trap, (6): Silica gel dryer for countercurrent air that flows into the Nafion dryer, (7): Inlet filter for ambient air; (8): four calibration gases (zero, 250, 350, 450 ppm); NDIR: Non-dispersive Infrared detector (LiCOR 6252).

The seawater flows into the equilibrator where thermodynamic equilibrium is reached with a non-circulating countercurrent flow of sample air (Figure 2). The tandem type equilibrator (Japanese Patent No. P2001-83053A) was designed by Y. Nojiri from the National Institute of Environmental Studies and Kimoto Inc (both Tsukuba, Japan). It consists of two stages: a bubbling equilibrator and static mixer equilibrator with a reported overall equilibration efficiency of 99.5% (Nojiri, personal communication). The equilibrated air is pumped into the measurement unit where it is dried in several steps. The first step includes a mechanical removal of droplets and condensate through a teflon filter (Figure 2: (1); Whatman polyflon filter, PF020, 47 mm) followed by a mist trap (2) and peltier element (3) operated at a dewpoint temperature of about 2°C. The air sample then flows into a Permapure® drier (4) where it is further dried using a countercurrent flow of dry ambient air. The latter is produced by a gas generator which consists of a pump and a dryer cylinder (6). The latter is filled with large granule silica gel that dries ambient air which is sucked in through an air filter (7). In the last drying stage the air is pumped through a magnesium perchlorate trap (5) which acts mainly as an indicator for successful drying. After drying the sample gas enters the detector (NDIR) unit (model 6252, LICOR, inc., Lincoln, USA). The CO_2 content of the gas is detected using a non-dispersive infrared method which has proven to be a robust and seaworthy method (Wanninkhof and Thoning, 1993). The gas measurement is calibrated against a suite of three calibration gases (working standards) with a nominal range of 250 to 450 ppmv plus CO_2 free nitrogen gas (8). The three standard gases were provided by Deuste & Steininger, Mühlhausen, Germany. The working standard gases were calibrated against NOAA primary standard gases of similar concentration range using a LICOR® 6262. This procedure yields an accuracy of 0.07 ppmv for our working standards relative to the NOAA corrected values.

The CO_2 concentration is calculated from the LICOR® mV signal. The latter is corrected for atmospheric pressure and cell temperature changes within the LICOR® unit. The corrected mV-readings are converted into CO_2 mixing ratios (xCO_2) using a least squares procedure for the quadratic regression function calculated from the calibration standards. The pCO_2 was calculated from the xCO_2 for 100% humidity and *in situ* temperature (remote sensor). The temperature deviation between the equilibrator and the *in situ* temperature was typically on the order of 0.01 to 0.03°C.

2.2. Air-Sea flux Calculation Schemes

For the calculation of the CO_2 air-sea flux density F [mmol m^{-2} d^{-1}] we used the following equation:

$$F (CO_2) = k\, K_0\, (pCO_{2\,sw} - pCO_{2\,atm}) \qquad (1)$$

where k is the transfer velocity, K_0 is the solubility of CO_2, and $pCO_{2\,sw}$ and $pCO_{2\,atm}$ are the partial pressures of CO_2 in seawater and atmosphere, respectively. The difference of the partial pressure of CO_2 between seawater and air (ΔpCO_2) is calculated from the observations. The transfer velocity k was calculated using four different parameterizations based on wind speed (Liss and Merlivat, 1986; Wanninkhof, 1992; Wanninkhof and McGillis, 1999; Nightingale et al., 2000). The solubility of CO_2 (K_0) was calculated by the equation of Weiss (1974) and the Schmidt number was calculated according to Wanninkhof (1992). Wind speed data were obtained from the 6-hourly NCEP/NCAR Reanalysis data provided by the NOAA-CIRES Climate Diagnostics Center, Boulder, Colorado, USA, (http://www.cdc.noaa.gov/). We also used climatological wind speed data obtained from COADS (Comprehensive Ocean Atmosphere Data Set).

We retrieved two ΔpCO_2 datasets based on different interpolation schemes. In dataset 1 the observations were averaged to a 1° x 1° grid which allows regional details of the CO_2 flux in the North Atlantic ocean to be identified (high resolution). In dataset 2 the observations are averaged to the 4°x5° grid of Takahashi et al. (2002) within the investigated area in order to compare with prior estimates of the air-sea CO_2 fluxes in this region (low resolution).

Note that we discriminate between CO_2 flux density (unit: mmol m^{-2} d^{-1}), annual CO_2 flux (unit: mol m^{-2} yr^{-1}) and annual oceanic CO_2 uptake (unit: g C yr^{-1}).

3. Results

3.1. Maps of ΔpCO_2 and CO_2 flux (High Resolution Dataset)

The annual cycle of ΔpCO_2 between February, 2002 and January, 2003 is presented in Figure 3 at a grid resolution of 1°x1° (high resolution). For most of the year this region of the North Atlantic Ocean is a potential sink for CO_2 and the annual mean ΔpCO_2 across the whole region is −28 ($\pm$ 14) µatm. Generally the disequilibrium is the largest during the spring

The variability of surface pCO₂ in the North Atlantic Ocean by Heike Lueger

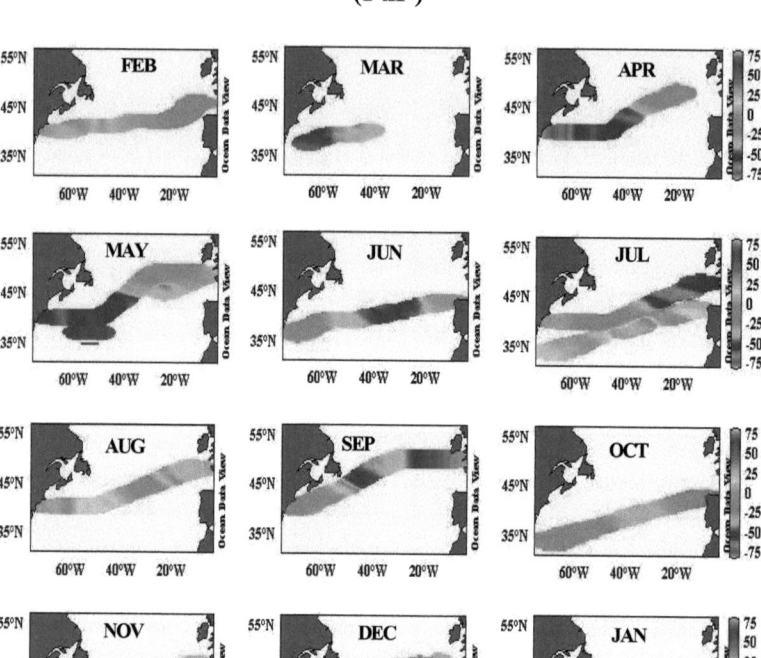

Figure 3 : Distribution of seawater ΔpCO_2 ($=pCO_{2\,seawater} - pCO_{2\,atmosphere}$) over the North Atlantic Ocean. Shown are the 1°x 1° averages of the observations between 2002 and 2003. Note that in some months there was more than one cruise.

months (March – June) when the seawater pCO_2 decreases due to biological drawdown. The spring bloom and resulting drawdown shows a temporal propagation into northeasterly direction with the strongest undersaturation found in the eastern basin during July (< -50µatm). During summer the seawater pCO_2 responds to the temperature rise and thus

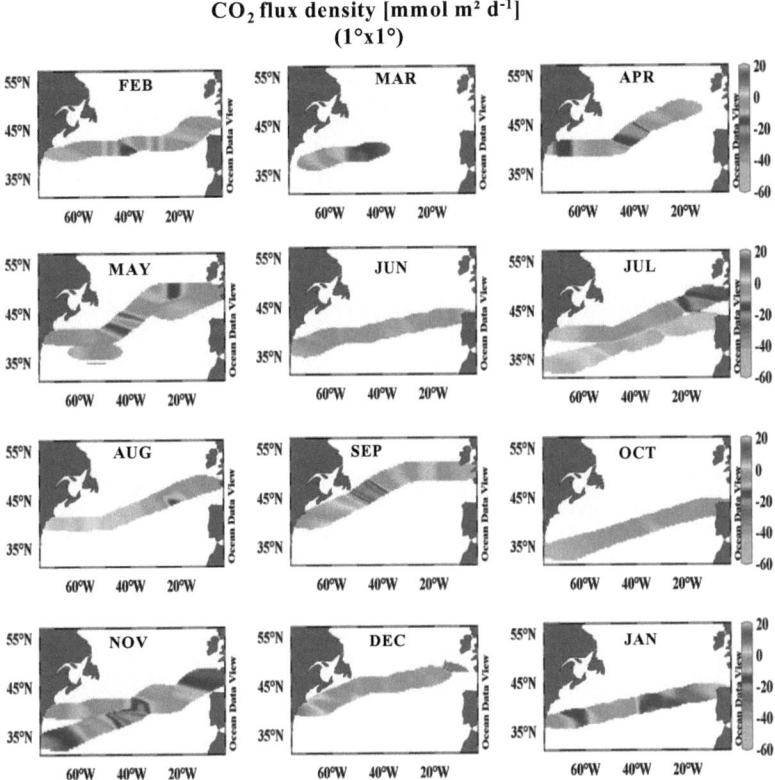

Figure 4: Distribution of CO_2 flux densities over the North Atlantic Ocean. Shown are the 1°x1° averages of the observations between 2002 and 2003. The CO_2 flux density is calculated from the ΔpCO_2 data and a gas transfer parameterization by Wanninkhof (1992) using 6 hourly wind speed data from NCEP/NCAR. Note that a negative sign denotes a flux from the atmosphere into the ocean.

reduces this disequilibrium. Only in July and August and in the western basin do we observe supersaturation of seawater pCO_2 with respect to the atmosphere ($\Delta pCO_2 > 0$ μatm).

We calculated the CO_2 flux density from ΔpCO_2 data and contemporary NCEP/NCAR winds (1°x1° dataset) using the gas transfer parameterization of Wanninkhof (1992) as described above. Generally, the CO_2 flux density mirrors the ΔpCO_2 (Figure 4). We find that this region of the North Atlantic is a net sink for atmospheric CO_2 with an annual mean CO_2 flux density of -6 ($\pm$ 8) mmol m^{-2} d^{-1}. Only in July and August do we observe a CO_2 source, with the highest values in the western basin during July (+ 7 mmol m^{-2} d^{-1}). In contrast to the ΔpCO_2 distribution, however, the strongest CO_2 sink is found during the winter season (-12 mmol m^{-2} d^{-1}) due to the effect of wind speed on the CO_2 flux density. Two strong but regionally restricted CO_2 sink areas can be identified over the annual cycle. In September the ship sailed through Labrador Current waters with lower temperature and salinity signature as well as lower pCO_2 values giving a CO_2 flux density minimum of –43 mmol m^{-2} d^{-1}, i.e. strong sink. In November a strong CO_2 sink at ~ 40°W (-27 mmol m^{-2} d^{-1}) can be explained primarily by high wind speeds (>15 m/s).

Three parameters contribute to calculations of CO_2 flux density: ΔpCO_2, wind speed, and the solubility of CO_2 (K_0) and all are shown in Figure 5. There is a distinct difference in ΔpCO_2 between the eastern (0-32°W) and the western (33-70°W) basins which is obvious when comparing monthly means of each basin computed from the 1°x1° dataset. In the western basin the maximum (positive) ΔpCO_2 is found in August (Figure 5a). In the eastern basin, however, the effect of temperature forcing during summer is damped by the counteracting effects of biology and air-sea exchange leading to lower (always negative) values of seawater pCO_2. These results – effects of temperature, biology, and air-sea gas exchange on the pCO_2 - are discussed in more detail elsewhere (Lueger et al., submitted). The eastern and western basin show a different wind speed pattern, too. In the western basin the wind speed is more uniformly distributed with a wind speed range of 6-11 ms^{-1} and an annual average wind speed of 8 ms^{-1}. In the eastern basin the wind speed shows a distinct seasonality with slightly higher annual average winds (9 ms^{-1}) than in the western basin and a greater range (6 to 14 ms^{-1}). K_0 mainly depends on temperature and the annual cycle shows lower values for the eastern basin than for the western basin (Figure 5c). The eastern and western basin show different annual temperature means of 15°C and 19°C, respectively, which explains the difference in K_0. The CO_2 flux density in the western basin mirrors mainly the ΔpCO_2 cycle. In the eastern basin wind speed contributes as much as the ΔpCO_2 to the CO_2 flux variability. The effect of solubility or temperature dependency (K_0) on the variability of the CO_2 flux is smaller in both basins than that of wind speed and ΔpCO_2.

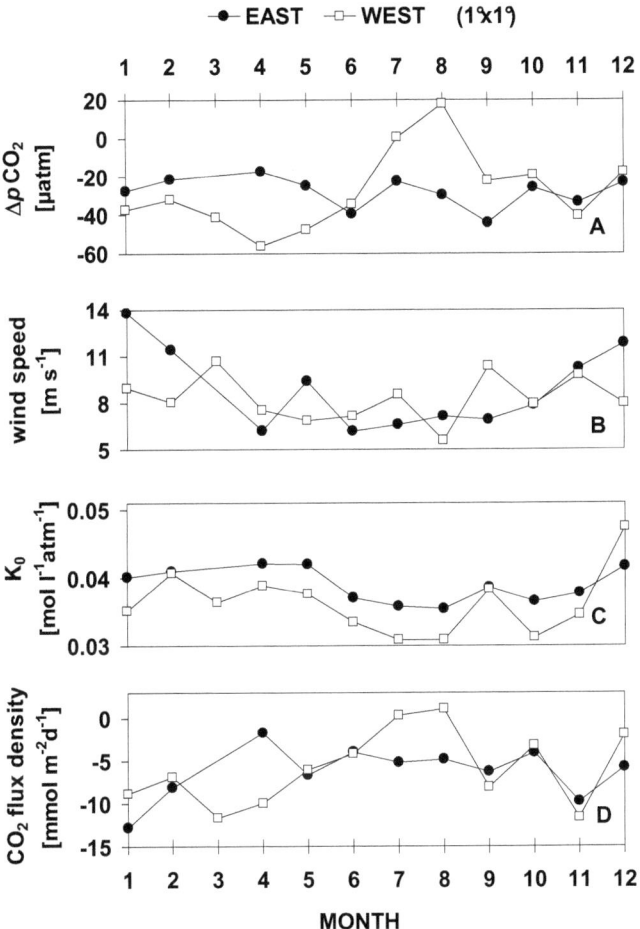

Figure 5 : Distribution of monthly means from the 1°x1° dataset of ΔpCO_2(A), wind speed (B), solubility coefficient K_0 (C), and CO_2 flux density (D) in the eastern (0°-32°W; black circles) and western (33°-70°W; white squares) basin. The CO_2 flux was calculated from the individual ΔpCO_2 data using the W92 parameterization for short-term winds and then averaged to monthly means. Wind speed data source: NCEP/NCAR.

3.2. Effect of spatial resolution on the CO_2 flux

The annual CO_2 flux density cycle of the 4°x5° grid is a coarse resolution and therefore it does not resolve small scale characteristics such as the November feature discussed above (Figure 6). However, the mean CO_2 flux – i.e. the annual average of all grid cells of the 1°x1

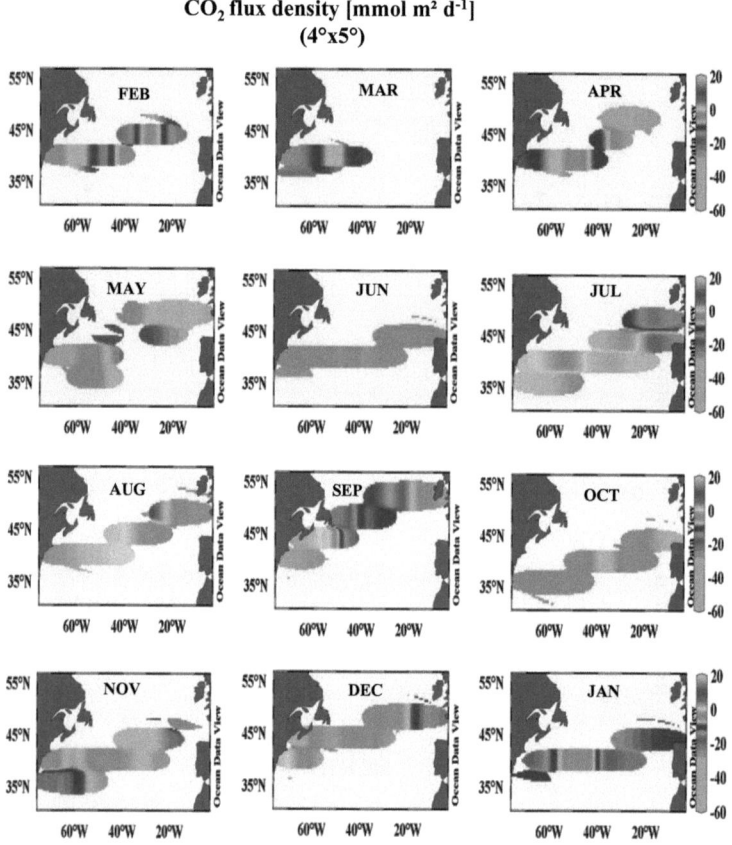

Figure 6: Distribution of CO_2 flux densities over the North Atlantic Ocean. Shown are the 4°x5° averages of the observations between 2002 and 2003. The CO_2 flux is calculated from the ΔpCO_2 data and a gas transfer parameterization by Wanninkhof (1992) using 6 hourly wind speed data from NCEP/NCAR. Note that a negative sign denotes a flux from the atmosphere into the ocean.

and 4°x5° datasets - is the same for both (–6 mmol m^{-2} d^{-1}), but the extreme values are shifted by the

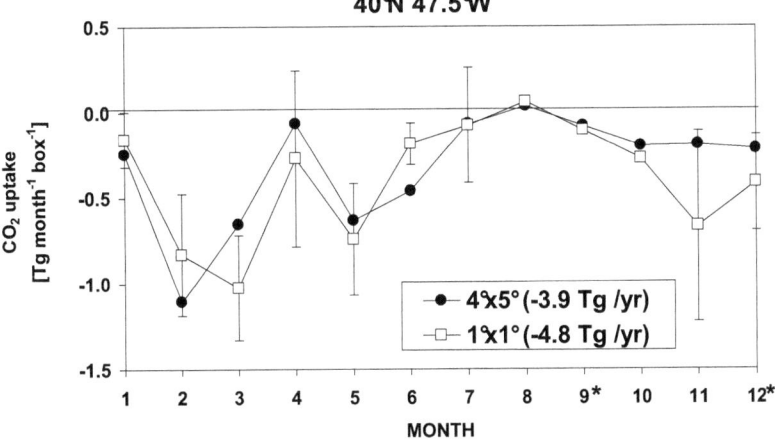

Figure 7: Comparison of CO$_2$ uptake rates (monthly means) between 1°x1° and 4°x5° resolution for one grid box (38-42°N, 45-50°W). The monthly uptake rates were calculated as follows: (a) 4°x5° gridding: the ΔpCO_2 values and winds (NCEP/NCAR) from the original dataset were averaged to this grid and then we calculated the monthly CO$_2$ flux which in turn was integrated over the box area and yielded the uptake rate; (b) 1°x1° gridding: the ΔpCO_2 values and winds (NCEP/NCAR) from the original dataset were first averaged to the 1°x1° grid, secondly to the 4°x5° grid, then the monthly CO$_2$ flux was calculated which in turn was integrated over the box area and yielded the uptake rate. Also shown is the annual CO$_2$ uptake, i.e. the sum of the monthly uptake values. The asterisk indicates that an interpolated value was used when there was no cruise in this grid box for this month. (Tg = 10^{12} g).

different averaging. When comparing the monthly means of the CO$_2$ fluxes we find the strongest CO$_2$ sink in the 4°x5° dataset in January whereas at 1x1° resolution it occurs in March. The weakest CO$_2$ sink is observed in August for both datasets. We also find that regionally important sinks such as that seen in November with the 1°x1° dataset, are not reflected at all in the 4x5° dataset. On the other hand larger scale CO$_2$ sources in July and August are similar in both with respect to size and location.

In order to retrieve a direct comparison between the two datasets we averaged the 1°x1° CO$_2$ flux densities to a corresponding single 4°x5° grid box in the western basin. We chose this grid box (40°N, 47.5°W) as here the monthly data coverage was high enough to represent a seasonal cycle (Figure 1). Both datasets show similar seasonal cycles and the difference between high and low resolution is smaller during summer and higher during the rest of the year (Figure 7). The annual CO_2 uptake rate in this box for the 1°x1° subset is 20% higher (-4.8 Tg/yr) than the coarse resolution (-3.9 Tg/yr).

3.3. Effect of wind speed data and parameterizations on the CO_2 flux (Low Resolution Dataset)

As the CO_2 flux depends *inter alia* on the parameterization of the gas transfer velocity k we also tested four different approaches (Figure 8). There has been a long series of publications suggesting various non-linear relationships between k and wind speed (u) and the differences between these approaches are significant. We use the 4°x5° dataset with the 6-hourly NCEP/NCAR winds to calculate the CO_2 flux density and annual CO_2 uptake using four well-cited parameterizations: Liss and Merlivat (1986; hereafter referred to as LM86), Wanninkhof (1992; W92) using the short-term parameterization of k, Wanninkhof and McGillis (1999; WM99), and Nightingale et al. (2000; N00). As expected all four parameterizations show the same seasonal pattern in the monthly CO_2 flux density, but differ in magnitude which is obvious from the resulting annual CO_2 uptake (table 1). The lowest CO_2 flux density and uptake is yielded by LM86 is which can be explained by the lower wind speed dependency of k. The Wanninkhof formulation (W92) yields the highest flux density and the annual CO_2 uptake is higher by over 40 % compared to LM86. The cubic relationship of WM99 gives an annual CO_2 flux density 13% lower to W92. The most recent parameterization (N00) gives CO_2 flux densities between those of LM86- and W92-based estimates, which is also reflected in the annual CO_2 uptake. Clearly all parameterizations differ most at high wind speeds and agree best at low wind speeds.

Because wind speed plays such a crucial role for the air-sea gas exchange of CO_2 we now test the effect of using different wind speed data sources. We compare short-term winds (NCEP/NCAR) of the 4°x5° dataset to wind speeds derived from the COADS climatology (same resolution). We averaged the wind speeds of all 4°x5° grid cells and all months of both

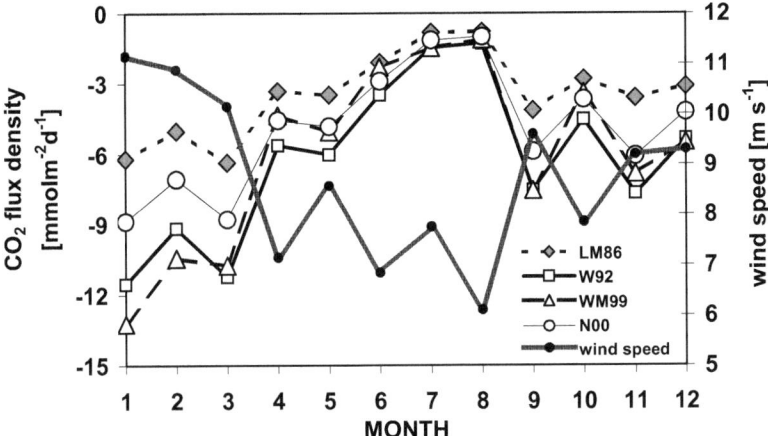

Figure 8: Comparison of the CO_2 flux densities (monthly means) calculated from four different parameterizations of the gas transfer velocity. LM86: Liss and Merlivat (1986), W92: Wanninkhof (1992), WM99: Wanninkhof and McGillis (1999), N20: Nightingale et al. (2000). Also shown are the monthly averaged wind speeds (courtesy NCEP/NCAR). We used the 4°x5° dataset.

Table 1: Annual CO_2 uptake summarized from the monthly means for four parameterizations in a grid box in the western basin. LM86: Liss and Merlivat (1986), W92: Wanninkhof (1992), WM99: Wanninkhof and McGillis (1999), N00: Nightingale et al. (2000). (1 Tg = 10^{12} gC).

40°N 47.5°W	LM86	W92	WM99	N00
annual CO_2 uptake [Tg C yr^{-1} grid cell^{-1}]	-2.2	-3.9	-3.4	-3.1

NCEP/NCAR and COADS data and found that the difference between the two is negligible (−0.01 m s^{-1}). The mean values are the same (8.5 m s^{-1}; Figure 9) which implies that 2002 was near the climatological mean in terms of wind speed. The climatological data, however, seem smoothed compared to the reanalysis data, but all differences are within the variance limits of the NCEP/NCAR data. Regional short-term wind events are of course not reflected in the climatology. For instance, in December at 48°N/ 12.5°W we find a very high difference between NCEP/NCAR and COADS wind speeds (7.5 m s^{-1}) which leads to an CO_2 sink that is increased by about 10 mmol m^{-2} d^{-1}(not shown).

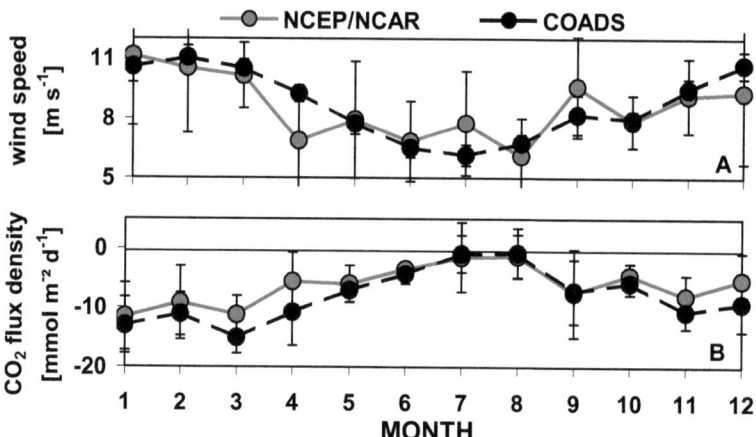

Figure 9: Comparison of the monthly means of wind speed data (A) and CO_2 flux densities (B) using different sources for wind speed: NCEP/NCAR reanalysis data (grey), COADS climatologoy (black). We used the 4°x5° dataset with the W92 parameterization for short-term winds (NCEP/NCAR) and climatological winds (COADS).

3.4. CO$_2$ flux estimates compared to a climatology (low resolution)

We calculated the CO$_2$ flux density of the Takahashi climatology using the climatological ΔpCO$_2$ values and the short-term (NCEP/NCAR) winds for 2002. Note that we only calculated CO$_2$ flux densities for the grid cells where we had data from our cruises (cf. Figure 1). Both 4°x5° datasets were first corrected to the virtual year 1995 to compensate for the anthropogenic impact (assuming a CO$_2$ increase of 1.5 ppm yr^{-1} for all grid boxes south of 45°N). The CO$_2$ flux densities were separated into different latitudinal bands (Figure 10).

The CO$_2$ flux densities of the most northern band (46-50°N) are shown in Figure 10a; data gaps indicate that there were no Falstaff data from this area at this time of the year. In this zonal band the CO$_2$ flux densities obtained from the Falstaff cruises and Takahashi's dataset agree during summer, but diverge in winter and spring. The CO$_2$ flux discrepancy in winter/spring can be as high as a factor of three. Next in the 42-46°N band both datasets reveal a year-round sink for atmospheric CO$_2$, although the Falstaff data suggest a very weak CO$_2$ source in July (+1 $\pm$ 4 mmol m^{-2} d^{-1}) which is not apparent in the climatology. In the 38-42°N band both datasets agree very well, except for September. In September Takahashi's climatological data imply a CO$_2$ source (5 mmol/m²/d), whereas the Falstaff observations suggest a CO$_2$ sink (-2 mmol m^{-2} d^{-1}). We attribute this to the influence of the Labrador current water in our dataset, as Takahashi's and Falstaff SSS data diverge for this period (Takahashi: 36.3; Falstaff: 34.6). When comparing the salinity data with the Levitus 94 climatology we find that both datasets diverge from the climatological mean SSS for this time and region (Levitus: 35.3). For the 34-38°N band we observe a larger bias between the datasets for most of the months, but it should be noted that our data coverage is severly reduced, too. Absence of error bars in January, March, and June indicate that only one 4°x5° datapoint was available for comparison in each of these months. Especially in June the difference between our data and the climatology is obvious: Takahashi's data suggests a source for atmospheric CO$_2$ (+8 mmol m^{-2} d^{-1}) whereas the Falstaff results indicate a weak CO$_2$ sink (-2 mmol m^{-2} d^{-1}). The difference is associated by the difference in salinity in June (Falstaff: 33.7, Takahashi: 36.1). This indicates that different water masses are considered in the two analyses. The salinity climatology by Levitus (Levitus 94) gives a value of 36.2 for this time and region which implies that the Falstaff data do not represent the climatological mean for this location and month.

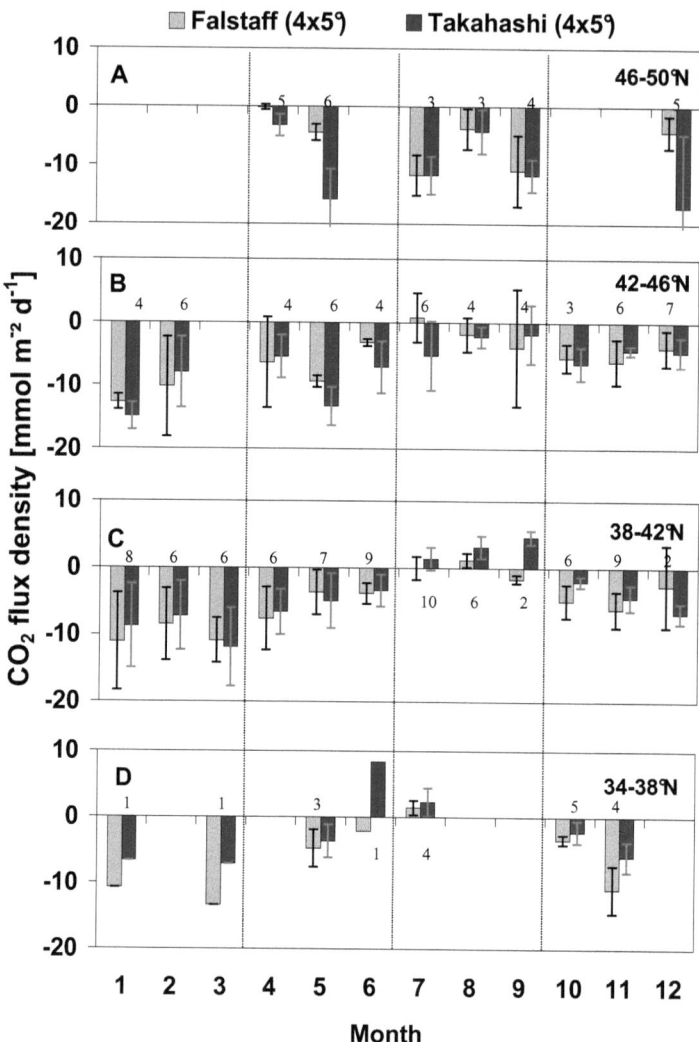

Figure 10: Comparison between the Falstaff and Takahashi et al. (2002) datasets using monthly mean CO_2 flux densities with short-term winds (NCEP/NCAR data). The CO_2 flux densities are monthly means and averaged for four latitudinal bands. The numbers above/below the bars indicate the number of datapoints (n) that were taken from the 4°x5° grid dataset and averaged for the means.

4. Discussion and Conclusions

We presented a full annual pCO_2 cycle of the latitudinal band (34°N-50°N) in the Atlantic Ocean. Independent gridding approaches yielded a similar result: this region of the North Atlantic ocean was a sink for CO_2 during most of 2002 with a mean annual CO_2 flux density for both 1°x1° and 4°x5° gridding methods over the entire area of 6 mmol m^{-2} d^{-1}. However, when comparing the high and low resolution we can state that generally large scale patterns are retained (e.g. summer CO_2 sources), whereas small scale features are damped (regional sinks) in the low resolution.

The seasonal cycles of seawater pCO_2 in the eastern and western basin in this region of the North Atlantic were analyzed in detail in a different manuscript (Lueger et al., submitted). In the latter we found that in the eastern basin the biological forcing on the pCO_2 was stronger (based on the correlation between pCO_2 and nitrate), whereas in the western basin the seawater pCO_2 primarily followed temperature changes. These patterns are generally reflected in the CO_2 flux when we correlate it to ΔpCO_2 and wind speed (Figure 11). The good correlation between CO_2 flux and ΔpCO_2 in the western basin is explained by the pronounced seasonal cycle of pCO_2 in the western basin. Here, the temperature effect on the seawater pCO_2 is not as strongly counteracted by biological forcing such as in the eastern basin resulting in a larger seasonal pCO_2 amplitude in the western basin. On the other hand wind speed has a greater impact on the CO_2 flux in the eastern basin than in the western basin. In Lueger et al. (submitted) we showed that the mixed layer depth reached much higher values in the eastern than in the western basin which also affects the supply of nutrients and subsequently increases the biological forcing.

The annual CO_2 uptake rate varied depending on which wind speed data were used (Table 2). The 4°x5° dataset using NCEP/NCAR (short-term) wind speed data yields an annual CO_2 uptake of 0.052 Pg C yr^{-1} which is 30% lower than the COADS (climatological) wind speeds dataset (0.071 Pg C yr^{-1}). Bates et al. (1998) stated that differences in wind speed data sets could be as high as 60-70% in the subtropical gyre of the North Atlantic. It is very important to accurately determine the wind speed whenever possible or to retrieve reliable high resolution estimates. However, it is also crucial to bear in mind that in many approaches different wind speed parameterization have been used. We showed that the annual CO_2 flux varied up to 40% depending on which gas transfer equation we used.

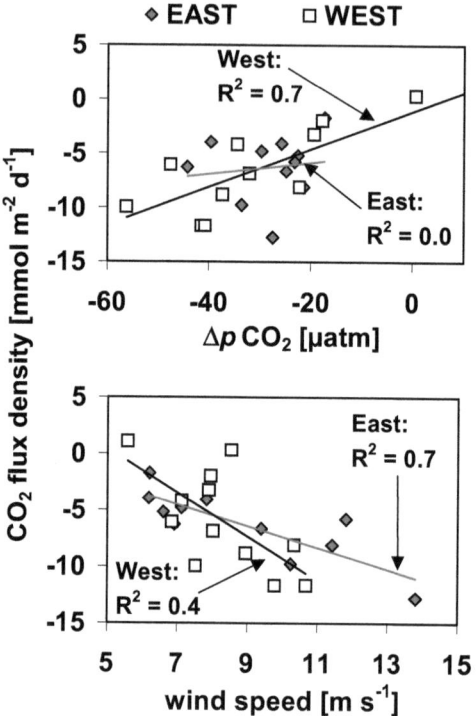

Figure 11: Comparison of eastern and western basin. The CO_2 flux densities were taken from the results presented in Figure 5 and plotted against the respective ΔpCO_2 (a) and wind (b) data. For more details please refer to figure 5.

Our annual oceanic CO_2 uptake results are comparable to results of other datasets, a direct comparison, however, is complicated by the fact that the examined regions differ drastically with respect to location, size, wind speed parameterization, and/or wind speed sources. Lefèvre et al. (1999) calculated the oceanic sink in the northern hemisphere for an extensive dataset and found an annual CO_2 uptake of -0.48 Pg C yr^{-1} (1.44 mol C m^{-2} yr^{-1}) for the North Atlanitc (10°N-50°N). Compared to our climatological dataset (COADS wind) this is within the same range (1.27 mol C m^{-2} yr^{-1}), but it is 40 % higher than our short-term estimate (0.93

mol C m^{-2} yr^{-1}). In the center of the subtropical gyre (32°N) Bates et al. (1998) found a CO_2 flux much lower (0.51 mol C yr^{-1}) than our results. On the

Table 2: Annual CO_2 uptake rates compared to published results. Our results were calculated from the 4°x5° dataset for the 38-46°N zonal band only as other regions did not have enough data coverage. Different parameterizations were used (W92: Wanninkhof (1992), T90: Tans et al. (1990).

	Falstaff	Falstaff	Takahashi (2002)	Lefèvre (1999)	Bates (1998)	Cooper (1998)
	4°x5°	4°x5°	4°x5°			
region	38-46°N	38-46°N	38-46°N	10-50°N	Sargasso Sea	15-45°N
Flux [Pg CO_2 yr^{-1}]	-0.052	-0.071	-0.054	-0.48	-	-
Flux [mol CO_2 m^{-2} yr^{-1}]	-0.93	-1.27	-0.97	-1.44	-0.51	-2.00
wind speed [m s^{-1}]	NCEP/NCAR	COADS*	NCEP/NCAR	SSMI*	daily	U.S. Navy*
Area [10^6 km²]	5	5	5	28	-	-
parameterization	W92 short-term	W92 climat.	W92 short-term	T90	W92 short-term	T90

*climatological wind speeds

other hand Cooper et al. (1998) yielded a CO_2 uptake rate in the subtropical gyre that is 30% (COADS) and over 50% (NCEP/NCAR) higher than our estimates.

When comparing the annual CO_2 uptake rate in the region between 38°N and 46°N between our dataset and Takahashi's climatology using the same wind speed and grid resolution, the overall difference is small (4 %; table 2). However, when computing the difference of the seasonally averaged CO_2 fluxes (table 3) it is obvious that the deviation between the two datasets is higher in winter (October to March) than in summer (April to September). This might be explained by a lack of wintertime data in the climatology. Usually most available ΔpCO_2 data are derived from research vessel cruises that generally avoid winter time due to difficult weather conditions. However, we do not have the sample density for Takahashi's climatology to analyze this in more detail. Furthermore, it is obvious that the deviations for the 48°N grid band are higher than in the other grid bands. The differences between the two datasets for the 48°N grid band decrease if we correct these data for the anthropogenic CO_2, i.e. if we assume an increase in CO_2 concentration in surface waters due to the rising atmospheric content (table 3). Another reason for the overall deviation between the two datasets could be, of course, that we only look at one year of data, and this year might well not be representative compared to the climatology.

This fairly high variability in CO_2 uptake rates is common and shows how important reliable estimates of the CO_2 flux seasonality are and that extensive sampling coverage is needed. It is obvious that consistent results can only be obtained when the scientific community agrees on a certain protocol regarding the choices of wind speed sources, perturbations (e.g. k) and corrections. Our results point out the importance of high-resolution sampling in the North Atlantic Ocean. We showed that it is possible to maintain a CO_2 measurement unit onboard commercial vessels and retrieve high quality data. We hope that CAVASSOO paved the way for future VOS projects in order to develop an extensive network of ships of opportunity. For now we will continue the VOS line in the future and contribute the data to international databases.

Table 3: Seasonal CO_2 flux differences between the Falstaff and Takahashi data calculated from the 3-monthly means of all four grid bands. Numbers in parenthesis (48°N band) show differences between CO_2 fluxes when all data are corrected to 1995. JFM: January-February-March, AMJ: April-March-June, JAS: July-August-September, OND: October-November-December. Mean denotes mean difference of all four grid bands.

mean CO_2 flux differences between Takahashi's climatology and Falstaff data						
Grid Band	36°N	40°N	44°N	48°N		mean 36°-48°N
JFM	5.2	0.9	0.0	-		2.0
AMJ	5.8	0.1	-2.3	-4.8	(-3.6)	-0.3
JAS	0.8	3.2	-1.4	-0.4	(2.1)	0.5
OND	3.1	0.2	0.1	-12.8	(-8.9)	-2.4

units are [mmol m^{-1} d^{-1}]

The variability of surface pCO$_2$ in the North Atlantic Ocean by Heike Lueger

Acknowledgements

We sincerely thank Per Croner and Sara Gorton of Wallenius Lines, Stockholm/Sweden, for cooperation and generous support. The outstanding support of the chief engineer, Gerth Gulliksson, as well as captain and crew of the M/V Falstaff is greatly appreciated. We also wish to acknowledge Yukihiro Nojiri for generously providing his automated pCO$_2$ measurement system. We thank Alexander Sy and the German Federal Maritime Agency (BSH) for providing the thermosalinograph. And thank to Peter Fritsche, Kerstin Nachtigall, Hergen Johannsen for chlorophyll and nutrient analysis, Lisa Weber, Philip Nuss, Tobias Steinhoff and Axel Wendt for cruise help, Hans-Peter Hansen for software support, Hela Mehrtens and Thomas Martin for help with the NCEP/NCAR wind speed data. This work was funded by the European Commission under grant no. EVK2-CT-2000-00088.

References

Bates, N.R., Takahashi, T., Chipman, D.W., Knap, A.H., 1998. Variability of pCO_2 on diel to seasonal timescales in the Sargasso Sea near Bermuda. *Journal of Geophysical Research, 103*, 15,567-15,858.

Cooper, D. J., Watson, A. J., Ling, R. D. 1998. Variation of pCO_2 along a North Atlantic shipping route (U.K. to the Caribbean): A year of automated observations. *Marine Chemistry, 60*, 147-164.

Dickson, R., Lazier, J., Meincke, J., Rhines, P., Swift, J., 1996. Long-term coordinated changes in the convective activity of the North Atlantic. *Prog. Oceanog., 38*, 241-295.

Lefèvre, N., Watson, A.J., Cooper, D.J., Weiss, R.F., Takahashi, T., Sutherland, S.C., 1999. Assessing the seasonality of the oceanic sink for CO_2 in the northern hemisphere. *Global Biogeochemical Cycles, 13(2)*, 273-286.

Liss, P.S., Merlivat, L., 1986. Air-sea gas exchange rates: Introduction and systhesis, in: *The Role of Air-Sea Exchange in Geochemical Cycling*, edited by P. Buat-Menard, Reidel, Boston, 113-129.

Nightingale, P.D., Malin, G., Law, C.S., Watson, A.J., Liss, P.S., Liddicoat, M.I., Boutin, J., Upstill-Goddard, R.C., 2000. In situ evaluation of air-sea gas exchange parameterizations using novel conservative and volatile tracers. *Global Biogeochemical cycles, 14 (1)*, 373-387.

Takahashi, T., Sutherland, S.C., Sweeney, C., Poisson, A., Metzl, N., Tilbrook, B., Bates, N., Wanninkhof, R., Feely, R., Sabine, C., Olafsson, J., Nojiri, Y., 2002. Global Sea-Air CO_2 flux based on climatological surface ocean pCO_2 and seasonal biological and temperature effects. *Deep Sea Research II, 49 (9-10)*, 1601-1622.

Tans, P., Fung, I.Y., Takahashi, T., 1990. Observational Constraints on the Global Atmospheric CO_2 Budget. *Science, 247*, 1431-1438.

Wallace, D.W.R., 2001. Storage and Transport of excess CO_2 in the Oceans: The JGOFS/WOCE Global CO_2 Survey. In: *Ocean Circulation and Climate*, 489-521.

Wanninkhof, R., 1992. Relationship between gas exchange and wind speed over the ocean. *Journal of Geophysical Research, 97*, 7373-7381.

Wanninkhof, R., Thoning, K., 1993. Measurement fo fugacity of CO_2 in surface water using continuous and discrete sampling methods. *Marine Chemistry, 44*, 189-204.

Wanninkhof, R., McGillis, W.R., 1999. A cubic relationhip between air-sea CO_2 exchange and wind speed. *Geophysical Research Letters, 26 (13)*, 1889-1892.

Weiss, R.F., 1974. Carbon dioxide in water and seawater: The solubility of a non-ideal gas. *Marine Chemistry, 2*, 203-215.

Zeng, J., Nojiri, Y., Murphy, P.P., Wong, C.S., Fujinuma, Y., 2002. a comparison of $\Delta p CO_2$ distributions in the northern North Pacific using results from a commerical vessel in 1995-1999. *Deep-Sea Research II, 49*, 5303-5315.

The variability of surface pCO₂ in the North Atlantic Ocean by Heike Lueger

Seasonal Cycles of Nutrients in the North Atlantic

Heike Lüger*

(corresponding author)

Douglas W.R. Wallace*

Arne Körtzinger*

Yukihiro Nojiri[§]

*Institut für Meereskunde an der Christian-Albrecht-Universität zu Kiel, Düsternbrooker Weg 20, 24105 Kiel, Germany

[§]National Institute for Environmental Studies, 16-2, Onogawa, Tsukuba, Ibaraki, 305-0053, Japan

Abstract

We present seasonal changes of nutrients, dissolved inorganic carbon (DIC), total alkalinity and chlorophyll a that were collected over one year in the mid-latitude North Atlantic onboard a commercial vessel. We divided the data into different cluster according to their T-S signatures: North Atlantic Drift (NADR), Gulf Stream (GSTR) and Labrador Current (LBRC). In the western provinces (GSTR, LBRC) the seasonal changes of nitrate are less pronounced compared to the eastern province (NADR) and the normalized DIC values show considerable differences between the provinces. During spring time we found a decrease in alkalinity concentrations for the NADR regime and showed that it could be attributed to calcification which increased the seasonal DIC loss by about 12%. The C:N ratio of the new production was highest for the GSTR (15.4) province accompanied by elevated N:P ratio (28.2) which indicates the strong influence of N_2 fixation in this region.

The variability of surface pCO₂ in the North Atlantic Ocean by Heike Lueger

1. **Introduction**

Biogeochemical cycling of nutrients in the ocean is complex and patterns of biological production are not well constrained. These processes are important for the oceanic carbon cycle as they modulate the CO_2 chemistry and subsequently the CO_2 sequestration. In the deep ocean nutrients are less affected by seasonal changes and mostly in steady-state conditions compared to the strong seasonality in the upper ocean (Glover and Brewer 1988). Biological activity continuously changes the chemistry of nutrients in the euphotic zone through photosynthesis and respiration the balance of which represents the net community production. New production is that part of the primary production which is based on allochthonous nutrients, i.e. nutrients imported into either from below or above the euphotic zone (Dugdale and Goering 1967). It can be estimated from the seasonal carbon and nitrate drawdown (end of winter – end of productive season); on an annual basis this should equal the export of organic matter which is part of the biological pump. Here, the choice of the wintertime reference point is crucial. In the North Atlantic ocean wintertime data of nutrients, however, are sparse as data sampling is mostly restricted to fair weather times. Climatologies of nutrients are often biased in this regard and therefore much work has been done to derive wintertime nutrients from correlated parameters. For example, Glover and Brewer (1988) extrapolated nutrient meaurements during spring and summer to the wintertime whereas Koeve (2001) and Körtzinger et al. (2001) used oxygen saturation to derive wintertime nutrients. We gathered a complete annual cycle of nutrient data in the mid-latitude North Atlantic and our wintertime data will be compared to the above approaches. We also present the seasonal cycle of nutrients and chlorophyll and calculate the seasonal drawdown of carbon and nitrogen. Further we report changes in total alkalinity that are caused by calcification during spring time in the Northeast Atlantic.

2. **Methods**

2.1. **Data collection**

In January 2002 the car carrier M/V Falstaff of the Swedish shipping company Wallenius Lines was equipped for shipboard pCO_2 measurements. Additional discrete samples were collected on board during 10 trans-Atlantic cruises between February 2002 and February

The variability of surface pCO$_2$ in the North Atlantic Ocean by Heike Lueger

2003. Discrete samples were taken every 3 to 6 hours except during night-time. Every cruise (Southampton – New York) took between 8 and 10 days and a roundtrip lasted about 42 days. The seawater intake was installed in the lowest deck of the engine room, approximately 8 m below the waterline. The seawater was piped into the pCO$_2$ system and a tubing was branched

Table 1: Cruises and discrete samples taken on board M/V Falstaff. The samples comprise nutrients, dissolved inorganic carbon, total alkalinity, and chlorophyll a. All samples were analyzed at the shore based laboratory.

Cruise #	time	Max Lat	Min Lat	Min Lon	Max Lon	samples (discrete)		
						Nuts	DIC/TA	Chl
FAL01	Feb, 2002	50°N	41°N	2°W	67°W	33	33	33
FAL03	Apr, 2002	49°N	41°N	20°W	71°W	28	28	28
FAL05	May, 2002	50°N	41°N	5°W	71°W	32	32	32
FAL07	Jul, 2002	50°N	41°N	5°W	70°W	52	31	31
FAL09	Aug, 2002	50°N	41°N	5°W	73°W	56	32	32
FAL11	Sep, 2002	50°N	41°N	4°W	73°W	44	26	26
FAL13	Nov, 2002	51°N	40°N	1°W	74°W	56	32	32
FAL15	Dec, 2002	50°N	41°N	5°W	70°W	49	29	29
FAL16	Jan, 2003	44°N	35°N	10°W	75°W	60	39	39
FAL17	Feb, 2003	50°N	41°N	6°W	69°W	46	30	30

off for collecting additional samples. The latter included nutrients, chlorophyll a, dissolved inorganic carbon (DIC), and total alkalinity (TA) samples. All samples (except DIC/TA) were stored onboard in a freezer (-20°C) for later shore based laboratory analysis. From February, 2002, to February, 2003, about 450 nutrient and 310 DIC / TA samples were collected (Table 1).

The nutrient samples were analyzed following the method of Hansen and Koroleff (1999). Chlorophyll samples (1-2 litres) were filtered onboard using glass fiber filters (GF/F) which were also stored in a freezer until lab analysis. Chlorophyll a was determined using the spectrophotometric method by Jeffrey and Humphrey (1975). The DIC and TA samples were preserved with 0.02 % HgCl$_2$ as described in the CDIAC handbook (Dickson and Goyet,

1994). DIC samples were determined following the coulometric technique of Johnson et al. (1999) using the SOMMA system. TA samples were analyzed by the potentiometric titration in an open cell as described in Mintrop et al. (2000). Both DIC and TA samples were referenced against seawater samples that were provided by A. Dickson (Scripps Institution of Oceanography, La Jolla, USA). The precision (accuracy) were for the DIC analysis ± 2 (± 1) μmol kg^{-1} and for the TA analysis ± 2 (± 2) μmol kg^{-1}.

The seasonality of nitrate* (nitrate + nitrite), normalized DIC and chlorophyll were analyzed with a sigmoidal function which can be expressed by a harmonic equation (after Zeng et al., 2002):

$$x(t) = c_0 + c_1 \sin(2\pi t) + c_2 \cos(2\pi t) + c_3 \sin(4\pi t) + c_4 \cos(4\pi t) \qquad (1)$$

where x is the seasonally varying quantity (e.g. nitrate*) and t is time (t = month). It is required that for the seasonal coverage the maximum data gap should not exceed three months.

2.2. Definition of Water Masses

The variability of the nutrients cycles is strongly affected by surface water regimes. Therefore we divided our dataset into subsets of different watermasses. Temperature-salinity relationships of our data were compared to data of the climatological atlas by Levitus (1994) in order to define water masses. Annual minimum and maximum of T and S from the climatology helped to identify the upper and lower T and S boundaries for each water masses (Figure 1a). Based on this we found three obvious data cluster which we describe in the following. Data that fell outside these clusters are not considered in the following analysis.

We defined the North Atlantic Drift (NADR) as an eastward flow which shows a temperature and salinity range of 11°C to 21°C and 34.9 to 35.9, respectively (Figure 1a). The zonal margins are between 40°N and 51°N covering a meridional range of 3°W to 64°W (Figure 1b). It is separated from Gulf Stream current (GSTR) which is also an eastward flowing regime. The TS-signature of this water mass is 15°C – 27°C and 36.0 – 36.6. The GSTR flows between 35°N - 40°N and 25°W - 75°W. The Labrador Current (LBRC) data display low temperature and salinity ranges (2°C – 10°C and 32 – 34, respectively). This watermass is primarily located in the western basin between 40°N - 46°N and 47°W – 70°W. Its origin lies in the Labrador Sea which is strongly affected by ice formation and melting.

The variability of surface pCO$_2$ in the North Atlantic Ocean by Heike Lueger

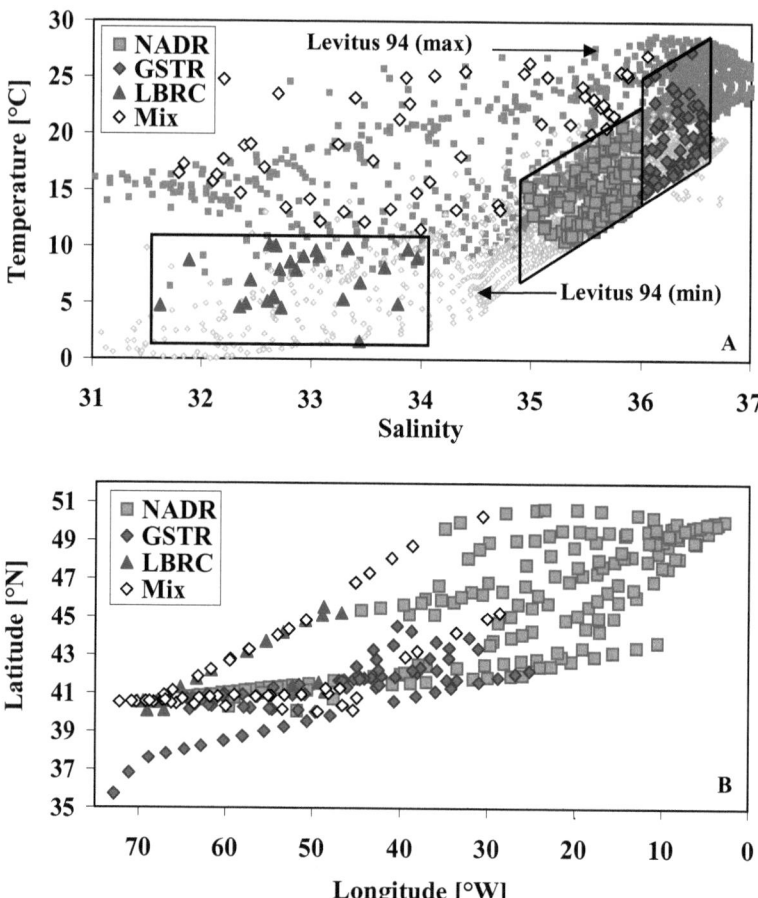

Figure 1: a -T-S diagram for all discrete samples between Feb. 2002 and Feb. 2003. The surface watermasses were identified based on their T-S-relationship and compared to the minimum and maximum T-S data of the Levitus climatology (grey). NADR - North Atlantic Drift, GSTR - Gulfstream, LBRC - Labrador Current, MIX - mixed watermass presumably containing NADR, GSTR, LBRC. **b** - geographical location of the such defined water masses.

2.3. Normalization of DIC and TA data

Normalization of dissolved inorganic carbon (DIC) and total alkalinity data (TA) to a constant salinity is a widespread practice. It helps to remove evaporation/precipitation effects that change the salinity and subsequently DIC and TA concentration. Both DIC and TA are normalized to a constant salinity following the approach proposed by Friis et al. (2003):

$$nX = ((X_{meas} - X_{S=0}) S_{meas}^{-1}) \times S_{ref} + X_{S=0} \qquad (2)$$

where X_{meas} denotes the observed DIC or TA and S_{meas} represents the salinity of the sample, $X_{S=0}$ is DIC/TA at zero salinity derived from a linear regression (see below), and S_{ref} is the mean salinity of the subset. We calculated the relationships between TA and salinity (SSS) for the three watermasses. Note that in the TA dataset we found a non-linear behaviour during the spring months and therefore we excluded these data. This deviation will be discussed later on.
In contrast the DIC concentration is more strongly affected by temperature than by salinity. Thus a similar approach as for the TA normalization is not possible as the DIC-SSS relationship is completely masked by temperature effects. Therefore we will calculate the linear regression of the DIC data from the TA-SSS relationship according to the following concept:

1. We define the linear regression for each watermass (NADR, GSTR, LBRC) from the TA-SSS relationship.
2. We normalize the TA data to a constant salinity using equation 1 and the individual linear regressions from 1.
3. The intercepts of the TA-SSS regressions (b_{TA}) for each watermass are considered as the DIC concentration at zero salinity (S_0)
4. We define the mean salinity (SSS_{mean}) and mean DIC (DIC_{mean}) values for each watermass.
5. All four parameters (b_{TA}, S_0, DIC_{mean}, SSS_{mean}) are used to calculate the linear regression for the DIC data of each watermass and finally to normalize the DIC data according to equation 1 (=nDIC).

In (3) we follow as a first order approximation the assumption that the DIC and TA river endmembers are equal. This assumption is based on the observation that bicarbonate ions constitute the vast majority of the dissolved inorganic carbon species in river water. A closer look at DIC and TA in river water do in fact differ according to the water's CO_2 partial pressure. The latter only has implications for CO_2 system calculations at low salinities (e.g. Körtzinger 2003), but is largely irrelevant for the normalization procedure followed here.

As mentioned above we observed a deviation of the linear relationship between TA and SSS. This is attributed to calcification, i.e. the formation of $CaCO_3$ by marine plankton. We determined the calcification term for NADR waters as only here we found a successive decrease in TA. We calculated the difference between observed and from SSS predicted TA. The difference equals the decrease in TA due to calcification and can be translated into a DIC change by dividing by a factor of 2.

2.4. Calculation of New Production

We estimate the seasonal new production (NP) from the depletion of nutrients during the productive season using nitrate (NP_N) and nDIC (NP_C) data based on the following scheme which was adapted from Körtzinger et al. (2001):

$$NP_N = NO_{3\,max} - NO_{3\,min} \qquad (3)$$

$$NP_C = nDIC_{max} - nDIC_{min} + ASE - C_{CaCO3} \qquad (4)$$

ASE denotes the DIC change due to air-sea exchange (see below) and C_{CaCO3} describes the effect of calcification on DIC.

DIC unlike nitrate is not only affected by biological processes but also by the air-sea exchange (ASE). We estimated the ASE effect from the net CO_2 flux which is derived from the difference of the partial pressure of CO_2:

$$F(CO_2) = k\,K_0\,(pCO_{2\,sw} - pCO_{2\,atm}) \qquad (5)$$

where k is the gas transport velocity, K_0 is the solubility of CO_2, $pCO_{2\,sw}$ and $_{atm}$ are the partial pressures of CO_2 of seawater and atmosphere, respectively. The pCO_2 values (seawater and atmosphere) were taken from continuous measurements. We calculated K_0 according to Weiss (1974) and k using the short-term parameterization of the wind speed by Wanninkhof (1992). The wind speed is obtained from NCEP/NCAR Reanalysis data provided by the NOAA-CIRES Climate Diagnostics Center, Boulder, Colorado, USA, (http://www.cdc.noaa.gov/). The mixed layer depth (MLD) data are taken from the Levitus 1994 climatology and the CO_2 flux data are divided by them which yields the cumulative DIC change due to the ASE effect. The ASE term is added to the salinity normalized DIC data (nDIC).

The variability of surface pCO$_2$ in the North Atlantic Ocean by Heike Lueger

3. Results

3.1. Wintertime nutrients

We compare our winter time observations of nutrients and CO$_2$ to winter estimates from earlier studies. The computed approaches tend to overestimate the nutrient concentration, but -

Table 2: Comparison of observed winter time nutrients with published estimates. The observed data were measured onboard the M/V Falstaff.

	Lat [°N]	Lon [°W]	NO$_3$ [µmol kg^{-1}]	PO$_4$ [µmol kg^{-1}]	Si(OH)$_4$ [µmol kg^{-1}]
observed winter max (April, 2002)[1]	49	20	7.3 ± 0.1	0.4 ± 0.01	2.8 ± 0.5
Koeve (2001)[2]	47	20	7.9 ± 0.3	0.5 ± 0.01	2.9 ± 0.1
Koeve (2001)[3]	47	20	8.1 ± 0.7	0.6 ± 0.05	3.2 ± 0.7
Körtzinger (2001)	46	20	6.6 ± 0.3	-	-
Körtzinger (2001)	49	20	8.4 ± 0.4	-	-
Glover and Brewer (1988)	47	20	12 ± 0.8	0.8 ± 0.8	6 ± 0.6
observed winter max (Feb, 2003)[1]	41	55	7.6 ± 0.1	0.6 ± 0.01	2 ± 0.5
observed winter max (Feb, 2002)[1]	41	55	3.9 ± 0.1	0.1 ± 0.01	5 ± 0.5
Glover and Brewer (1988)	40	55	4 ± 0.8	0.6 ± 0.8	2 ± 0.6

[1] errors (Falstaff): Methods of Seawater Analysis (1999)
[2] database: Atlantis II cruise 119/4
[3] database: Meteor curise 10/2

except for the estimate by Glover and Brewer (1988) - the prediction is very close to the observed value, often within the error bar limits (Table 2). The more recent estimates of wintertime nutrients (Koeve 2001; Körtzinger et al. 2001) are in better agreement than the earlier estimates by Glover and Brewer (1988) which display the highest deviations from the observed values (more than 50%). The mean deviation in nitrate, phosphate and silicate concentration between Koeve (2001) and our data is around $\pm$ 9 %, $\pm$ 25 %, and $\pm$ 8 %, respectively (Körtzinger et al. 2001: nitrate: $\pm$ 2 %). Also shown is an example of the interannual variability of the observed values in the western part of the North Atlantic (41°N and 55°W). In February, 2002 and 2003 the M/S Falstaff sailed nearly identical cruise tracks. All nutrient samples of this location differ interannually between 40% and 80% which is equivalent to the deviations between estimates of Glover and Brewer (1988) and our data.

3.2. Annual cycles of nutrients, chlorophyll a and carbon dioxide parameters

We present the annual cycles of nitrate and nitrite (=nitrate*), chlorophyll a, normalized dissolved inorganic carbon (nDIC) and total alkalinity (nTA) between February, 2002 and February, 2003. In all water masses minimum values are found during the summer months and maximal values occur during the spring or wintertime. The mean annual nitrate* concentration is highest in LBRC waters and lowest in GSTR waters (Figure 2). The NADR displays maximal and minimal nitrate* concentration in April (monthly mean: 5.9 μmol kg^{-1}) and August (0.1 μmol kg^{-1}), respectively. The data scatter is considerable in April and May which we attribute to the onset of the productive season that can be irregular and patchy. The GSTR nitrate data are more uniform than the NADR data, but also show a clear seasonal variation. Here maximum and minimum nitrate* values are found in February (monthly mean: 4.1 μmol kg^{-1}) and August (0.1 μmol kg^{-1}), respectively, which shows that the seasonal amplitude of nitrate is 30% lower than NADR waters. In the LBRC regime the seasonal amplitude of nitrate* is pronounced reaching maximum and minimum monthly mean values of 5 μmol kg^{-1} (February) and 0.1 μmol kg^{-1} (September), respectively. In December the nitrate concentrations are about 4 μmol kg^{-1} and therefore in the same range as NADR data. However, the nitrate* concentrations in February 2003 reach values up to 12 μmol kg^{-1} and therefore they are considerably higher than in February 2002. The maximum and minimum normalized DIC data of the NADR regime are found in May and August (Figure 2), respectively, yielding a seasonal amplitude of 49 μmol kg^{-1}. In the GSTR dataset the normalized DIC data show a maximum and minimum in February and August, respectively. The seasonal amplitude is in the same range (53 μmol kg^{-1}) as for the NADR regime. In LBRC waters the nDIC data and the seasonal amplitude (28 μmol kg^{-1}) are significantly lower than in the other two water masses.

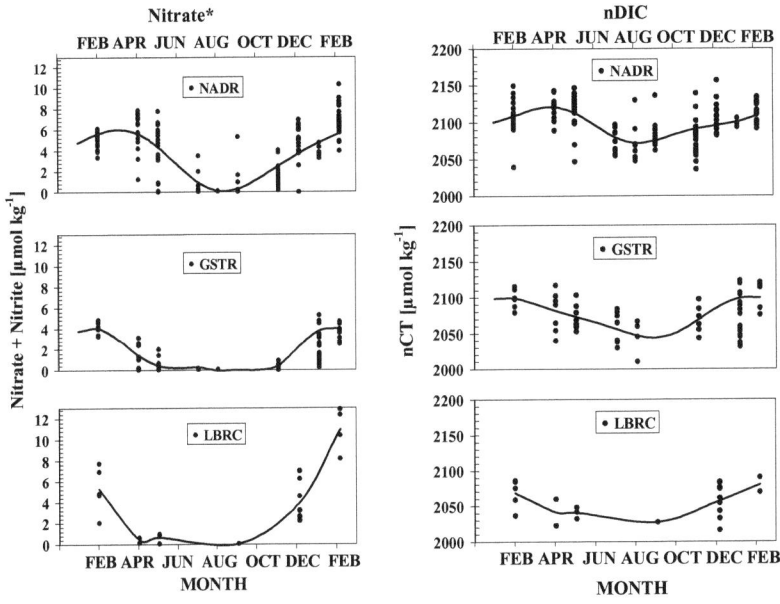

Figure 2: Seasonal cycles of nitrate* and normalized DIC in North Atlantic Drift (NADR), Gulfstream Current (GSTR), and Labrador Current water masses. Black lines indicate mean values computed from the following harmonic fit, except for LBRC where the monthly mean was used.

In the NADR chlorophyll concentrations are highest in April (monthly mean: 1.7 µg L^{-1}) reaching peak concentrations of over 5 µg L^{-1} (Figure 3). Between May and August chlorophyll a monthly mean values are between 0.6 and 1 µg L^{-1}. In August we found a very high chlorophyll concentration (4.6 µg L^{-1}) which goes along with decreased DIC and TA values (not shown). This datapoint which was found near the English channel (6.5°W) is likely to be influenced by coastal waters where biological productivity is higher. In November the chlorophyll a concentration rises up to 1.6 µg L^{-1} which points to a fall bloom. The seasonal cycle of GSTR chlorophyll a data shows two peak concentrations in April (1.8 µg L^{-1}) and November (0.6 µg L^{-1}) indicating a spring and a fall bloom. We do not find a decrease in nitrate* or nDIC data during fall which would go along with a bloom. On the other hand we also do not have samples for October where such a decrease could have occurred. The LBRC data show a pronounced peak in April which corresponds to the drawdown of nitrate. We do not find any signs of a fall bloom in this subset, which also may be due to the low data coverage.

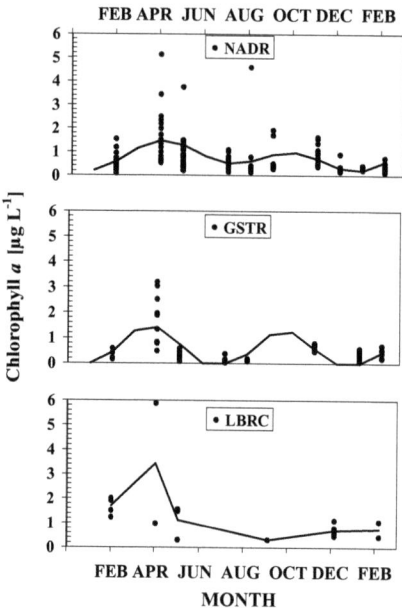

Figure 3: Seasonal cycles of chlorophyll in North Atlantic Drift (NADR), Gulfstream Current (GSTR), and Labrador Current water masses. Black lines indicate mean values computed from the following harmonic fit, except for LBRC where the monthly mean was used.

3.3. Calcification

In the North Atlantic total alkalinity (TA) is usually tightly correlated to salinity and therefore regarded as a conservative parameter (Millero 1998). We compare TA with salinity data for the different water masses (Figure 4a) and find a good correlation between SSS and TA (r^2= 0.9). Generally GSTR showed highest salinity and therefore highest TA values and LBRC data display lowest values in both SSS and TA. But we also observe a departure from the conservative behaviour in TA for the spring months (April – May) in NADR data (Figure 4b).

Some TA values deviate up to 34 µmol kg^{-1} from the lowest error margin (1σ). We attribute this feature to calcification of marine organisms. Coccolithophorids are able to form calcareous shells which can be remotely sensed due to the light scattering of calcite. We explored satellite images for the time and region where the TA drop occurred and found a pronounced coccolithophorid bloom (Figure 5). Therefore we attribute the TA changes to this group rather than to foraminifera or pteropods.

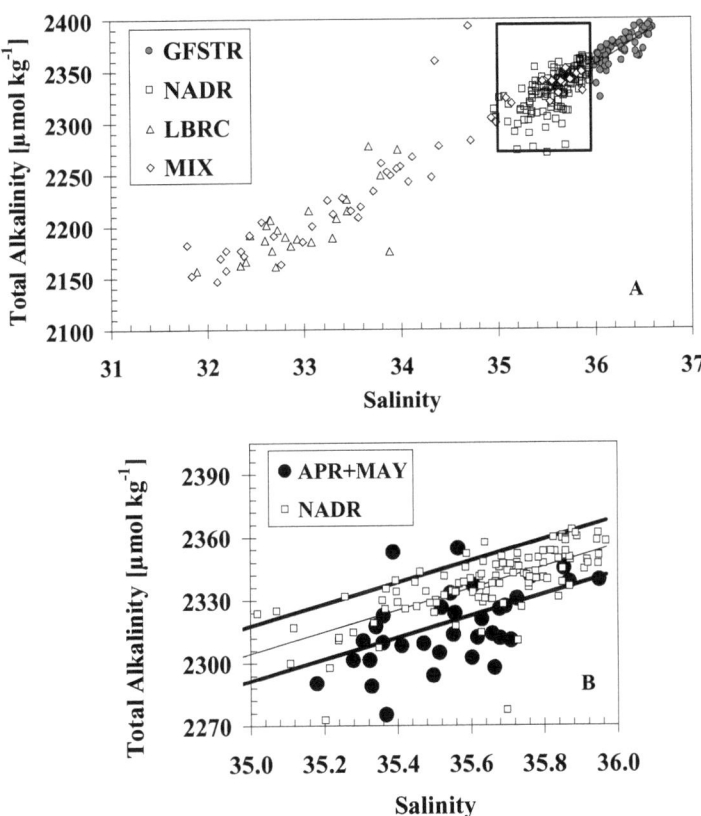

Figure 4: Relationship between total alkalinity (TA) and salinity in different water masses. **a**: The following regression equation is found for the dataset containing no NADR spring values: TA = 51.9 (±0.4) SSS + 490.9, r^2=0.98. **b** (cutout of figure a): SSS/TA relationships for April and May (black circles) cruises (NADR data only). The black lines indicate the TA-estimate from salinity data and the 1σ margins.

The variability of surface pCO$_2$ in the North Atlantic Ocean by Heike Lueger

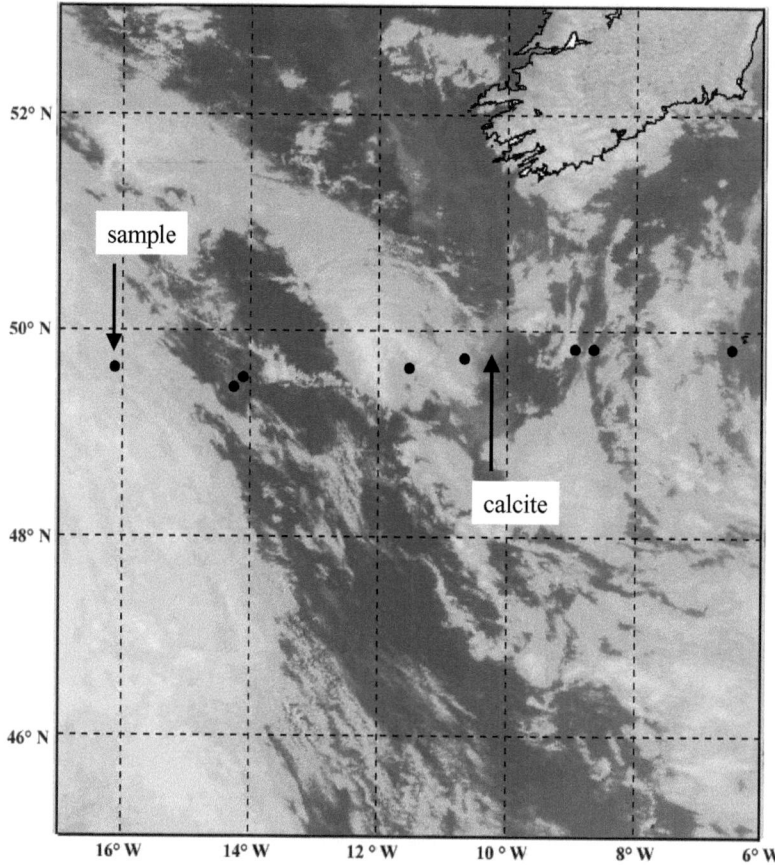

Figure 5: Satellite image of a coccolithophorid bloom in the North Atlantic in May, 2002. The black dots indicate some of the sample locations where we measured a significant drop in alkalinity (April and May). The satellite image was derived from the SeaWifs project.

3.4. New Production

We define the new carbon or nitrogen production as the net removal of DIC or nitrate, respectively, during the productive season (winter max – summer min). In contrast to nitrate, DIC is not only affected by biological processes (including calcification) but also by air-sea exchange of CO_2. The correction schemes were described above. The ASE effect slightly increases nDIC at the surface as the the CO_2 flux is always into the ocean. For GSTR and LBRC waters the ASE effect is similar and increases the carbon loss terms on the order of 14 % and 24 %, respectively. In NADR waters on the other hand it slightly decreases the carbon loss term by 6 %.

Table 3: Mean carbon (DIC), nitrogen and phosphorus loss terms for different hydrological regimes: NADR - North Atlantic Drift, GSTR - Gulf Stream Current, LBRC - Labrador Current. The loss terms were calculated from the difference between maximal (winter/spring) and minimal (summer) monthly mean concentrations of nDIC, nitrate and phosphate, respectively. nDIC data were corrected for the effect of air-sea exchange of CO_2 (all regimes) and calcification (NADR waters only).

	mean C loss [μmol kg^{-1}]	mean N loss [μmol kg^{-1}]	mean P loss [μmol kg^{-1}]	Winter Max	Summer Min	C:N ratio	N:P ratio
NADR	52.4	5.8	0.3	May	Aug	9.0	21.9
GSTR	61.4	4.0	0.1	Feb	Aug	15.4	28.2
LBRC	36.2	5.2	0.2	Feb	Sep	7.0	24.2

Calcification decreases the nDIC observations (NADR) for April and May in the North East Atlantic by 5 and 7 μmol kg^{-1}, respectively (TA decrease: 10 and 14 μmol kg^{-1}). The maximal nDIC concentration (ASE corrected) is found in May (2123.9 μmol kg^{-1}) and therefore we added the May calcification value of 7 μmol kg^{-1}. The carbon loss terms for all the three water regimes differ significantly (Table 3). The carbon loss for NADR waters (52.4 μmol kg^{-1}) is almost the same as the seasonal DIC amplitude as the ASE effect was counteracted by the correction for calcification on the biological drawdown. In the GSTR the biological drawdown is highest (61.4 μmol kg^{-1}) as the nDIC concentration is increased due to the ASE effect. The lowest carbon loss term is found for LBRC waters (36.2 μmol kg^{-1}). This is slightly higher than the seasonal DIC amplitude (27.6 μmol kg^{-1}) which is also due to the increasing effect of ASE. The nitrogen and phosphorus loss terms show a different pattern compared to the carbon loss terms decreasing from NADR, LBRC to GSTR. Another difference in biogeochemical cycles of the water masses is the duration of the biological drawdown. GSTR and LBRC water masses show an earlier start of nutrient depletion than NADR water masses. By August, however, all regimes reach minimum values in nitrate and DIC. These time values give only an estimate as we do not have complete coverage of the annual cycle, e.g. we do not have samples for the month of March.

Based on the calculation of the loss terms we calculated the C:N ratios ($[C:N]_{NP}$). In the GSTR watermasses $(C:N)_{NP}$ is nearly 60% higher than the Redfield ratio of 6.6 whereas NADR and LBRC display values $\pm$ 1 of the Redfield ratio (Table 3). The N:P ratio for the GSTR (28.2) is almost 50% higher than the Redfield ratio of 16. The N:P ratios for the NADR (21.9) and LBRC (24.2) regimes are lower, but still increased compared to Redfield.

4. Discussion

We showed that published estimates of wintertime nutrient concentrations agree fairly well with our measurements when computed from oxygen saturation. On the other hand extrapolating spring and summer nutrient concentrations to wintertime seems to be less reliable. This is an important finding as the spatiotemporal distribution of nutrients especially during the wintertime is poorly constrained and estimates of wintertime nutrients are still in demand. These results encourage both refining estimation procedures and extensive collection of wintertime nutrients. Increasing the nutrient sampling, e.g. with the help of volunteer observing ships, is also important to detect the interannual variability, which – as we showed – can be considerable.

Seasonal cycles of nutrients are mainly influenced by biological processes and vertical mixing or advection. In this context it is noteworthy that the mixed layer depth is considerably higher in the eastern than in the western North Atlantic which is discussed elsewhere (Lueger et al. submitted to Global Biogeochemical Cycles). The higher mixed layer depth increases the nutrient supply and the NADR regime is affected by this and therefore the seasonal amplitude of nitrate is higher than in GSTR and LBRC waters. It is important to add that the seasonal changes of chlorophyll are less informative. This is explained by the fact that chlorophyll a is a highly transient tracer and over the annual cycle it shows a low correlation with the biological drawdown of carbon and nitrate.

The variability of surface pCO_2 in the North Atlantic Ocean by Heike Lueger

We presented a deviation of TA from the conservative behaviour with salinity which can occur if calcium carbonate ($CaCO_3$) is precipitated (TA decrease) or dissolved (TA increase). Such departures from the conservative behaviour of TA were observed earlier. A prominent example is shown in Bates et al. (1996). They found a significant decrease in TA of up to 30 $\mu mol\ kg^{-1}$ and showed that it was caused by open-ocean calcification of carbonate-secreting organisms. This compares to our observations as we found a maximal TA deviation of 43 $\mu mol\ kg^{-1}$ (May). Sambrotto et al. (1993) quantify the calcification effect on the DIC pool in the same region lower than 10%, which supports our study - we found that calcification enhanced the DIC loss by about 12%. This important result shows that calcification plays a considerable role in the carbon cycle of the ocean and should not be neglected.

The overall variability of the C:N ratio of the DIC and nitrate drawdown ($[C:N]_{NP}$) was close or slightly above the Redfield ratio in the LBRC and NADR waters, respectively. $(C:N)_{NP}$ in NADR was 9 which is within the range of earlier studies (Körtzinger et al. 2001; Rees et al. 1999). In GSTR waters, however, the deviation from the Redfield value was significant and this higher DIC loss is generally referred to as carbon overconsumption. The term was suggested by Toggweiler (1993) and describes the discrepancy of prevailing carbon consumption after nitrate concentrations are depleted. One possible explanation for this can be N_2 fixation by marine organisms which provides an alternative nitrogen source to nitrate (Gruber and Sarmiento 1997; Karl et al. 2002). It could be shown previously that N_2 fixation plays an important role in the subtropical gyre of the North Atlantic (e.g. Lipschultz et al. 2002) and that one indicator for N_2 fixation is an N:P ratio higher than the Redfield ratio (Karl et al. 2002). We found the highest N:P ratios in the GSTR waters (28.2) which supports the assumption that N_2 fixation plays an important role in this regime.

5. Conclusions

We presented a dataset retrieved from a volunteer observing ship with an extensive surface data coverage for more than one year. For this kind of dataset it is cruical to consider oceanographical provinces. Using temperature and salinity signatures to define these provinces is a helpful means. We showed that seasonal cycles of nutrients in the North Atlantic differ significantly for different water masses. The distribution of seasonal new production shows distinct patterns which is imprinted on the C:N ratio. Carbon overconsumption is obvious in Gulf Stream waters and may be attributed to N_2 fixation as indicated by elevated N:P ratios. We also showed that such a dataset is suitable to resolve seasonal events such as coccolithophorid blooms which influence the carbon cycle through calcite precipitation.

Acknowledgements

We sincerly thank Per Croner and Sara Gorton of Wallenius Lines, Stockholm/Sweden, for cooperation and generous support. The outstanding support of the chief engineer, Gerth Gulliksson, as well as captain and crew of the M/V Falstaff is greatly appreciated. We also wish to acknowledge Yukihiro Nojiri for generously providing his automated pCO$_2$ measurement system. We thank Alexander Sy and the German Federal Maritime Agency (BSH) for providing the thermosalinograph. And thank to Peter Fritsche, Kerstin Nachtigall, Hergen Johannsen for chlorophyll and nutrient analysis, Lisa Weber, Philip Nuss, Tobias Steinhoff and Axel Wendt for cruise help, Hans-Peter Hansen for software support, Hela Mehrtens and Thomas Martin for help with the NCEP/NCAR wind speed data, and Alexander Davidov for extracting the Seawifs image. This work was funded by the European Commission under grant no. EVK2-CT-2000-00088.

References

Bates, N. R., A. F. Michaels, A. H. Knap. 1996. Alkalinity changes in the Sargasso Sea: Geochemical evidence of calcification? Mar. Chem. **51**: 347-358.

Dickson, A. G. and C. Goyet. 1994. Handbook of methods for the analysis of the various parameters of the carbon dioxide system in sea water. Version 2, Carbon Dioxide Information Analysis Center, Report ORNL/CDIAC-74, Oak Ridge National Laboratory, Oak Ridge, Tennessee, USA.

Donald, K. M., I. Joint, A. P. Rees, E. M. Woodward, G. Savidge. 2001. Uptake of carbon, nitrogen and phosphorus by phytoplankton along the 20°W meridian in the NE Atlantic between 57.5°N and 37°N. Deep-Sea Res. **48**: 873-897.

Dugdale, R. C., J. J. Goering. 1967. Uptake of new and regenerated forms of nitrogen in primary productivity. Limnol. Oceanog. **12**: 196-206.

Friis, K., A. Körtzinger, D. W. R. Wallace. 2003. The salinity normalization of marine inorganic carbon chemistry data. Geophysical Research Letters **30**: 57-1.

Glover, D. M., P. G. Brewer. 1988. Estimates of wintertime mixed layer nutrient concentrations in the North Atlantic. Deep-Sea Res. **35**: 1525-1546.

Gruber, N., J. L. Sarmiento. 1997. Global patterns of marine nitrogen fixation and denitrification. Glob. Biogeochem. Cycles **11**: 235-266.

Hansen, H. P., F. Koroleff. 1999. Determination of nutrients. p. 159-228. *In*: K. Grasshoff, K. Kremling, M. Ehrhardt [eds.], Methods of Seawater Analysis. Verlag Chemie. Weinheim.

Jeffrey, S. W., G. F. Humphrey. 1975. New spectrophotometric equations for determining chlorophylls a, b, c_1, and c_2 in higher plants, algae and natural phytoplankton. Biochemie und Physiologie der Pflanzen **167**: 191-198.

Johnson, K.M., A. Körtzinger, L. Mintrop, J. C.Duinker, D. W. R.Wallace. 1999. Coulometric total carbon dioxide analysis for marine studies: measurement and internal consistency of underway surface TCO_2 concentrations. Mar. Chem. **67**: 123-144.

Karl, D., A. Michaels, B. Bergman, D. Capone, E. Carpenter, R. Letelier, F. Lipschultz, H. Paerl, D. Sigman, L. Stal. 2002. Dinitrogen fixation in the world's oceans. Biogeochemistry **57/58**: 47-98.

Körtzinger, A., W. Koeve, P. Kähler, L. Mintrop. 2001. C:N ratios in the mixed layer during productive season in the northeast Atlantic Ocean. Deep-Sea Res. **48**: 661-688.

Koeve, W. 2001. Wintertime nutrients in the North Atlantic – new approaches and implications for new production estimates. Mar. Chem. **74**: 245-260.

Levitus, S., T. Boyer. 1994. World Ocean Atlas Volume 4: Temperature, NOAA Atlas NESDIS 4, U.S. Department of Commerce, Washington, D.C.

Lipschultz, F., N. R. Bates, C. A. Carlson, D. A. Hansell. 2002. New production in the Sargasso Sea: History and current status. Glob. Biogeochem. Cycles **16**: 1-17.

Longhurst, A., 1995. Seasonal cycles of pelagic production and consumption. Prog. Oceanog. **36**: 77-167.

Millero, F.J., K. Lee, M. Roche. 1998. Distribution of alkalinity in the surface waters of the major oceans. Mar. Chem. **60**: 111-130.

Mintrop, L., F. F. Pérez, M. Gonzalez-Davila, J. M. Santana-Casiano, A. Körtzinger. 2000. Alkalinity determination by potentiometry – intercalibration using three different methods. Ciencias Marinas **26**: 23-37.

Pilson, M. E. Q., 1998. An introduction to the chemistry of the sea. Prentice-Hall, Inc.

Redfield, A. C., B. H. Ketchum, F. A. Richards. 1963. The influence of organisms on the composition of sea water, p. 26-77 *In*: M. N. Hill [ed.], The Sea, Vol.2. Interscience, New York.

Rees, A. P., I. Joint, K. M., Donald. 1999. Early spring bloom phytoplankton-nutrient dynamics at the Celtic Sea Shelf Break. Deep-Sea Res. **46**: 483-510.

Sambrotto, R. N., G. Savidge, C. Robinson, P. Boyd, T. Takahashi, D. M. Karl, C. Langdon, D. Chipman, J. Marra, L. Codispoti. 1993. Elevated consumption of carbon relative to nitrogen in the surface ocean. Nature **363**: 248-250.

Sarmiento, J. L., R. Murnane, C. LeQuéré. 1995. Air-sea CO_2 transfer and the carbon budget of the North Atlantic. Philos. Trans. R. Soc. Lond. **348**: 211-219.

Toggweiler, J. R. 1993. Carbon overconsumption. Nature **363**: 210-211.

Tomczak, M., J. S. Godfrey. 2001. Regional oceanography: An introduction. Webpage: http://www.es.flinders.edu.au/~mattom/regoc/

Wanninkhof, R. 1992. Relationship between wind speed and gas exchange over the ocean. Journal of Geophysical Research **97**: 7373-7382.

Weiss, R. F. 1974. Carbon dioxide in water and seawater: The solubility of a non-ideal gas. Marine Chemistry **2**: 203-215.

Wong, C. S., N. A. D. Waser, Y. Nojiri, F. A.Whitney, J. S. Page, J. Zeng. 2002. Seasonal cycles of nutrients and dissolved inorganic carbon at high and mid latitudes in the North Pacific Ocean during the Skaugran cruises: determination of new production and nutrient uptake ratios. Deep-Sea Res. **49**: 5317-5338.

Zeebe, R. E., D. Wolf-Gladrow. 2001. CO_2 in Seawater: Equilibrium, kinetics, isotopes. Elsevier Science.

Zeng, J., Y. Nojiri, P. P. Murphy, C. S. Wong, Y. Fujinuma. 2002. A comparison of ΔpCO_2 distributions in the northern North Pacific using results from a commerical vessel in 1995-1999. Deep-Sea Res. II **49**: 5303-5315.

Die VDM Verlagsservicegesellschaft sucht für wissenschaftliche Verlage abgeschlossene und herausragende

Dissertationen, Habilitationen, Diplomarbeiten, Master Theses, Magisterarbeiten usw.

für die kostenlose Publikation als Fachbuch.

Sie verfügen über eine Arbeit, die hohen inhaltlichen und formalen Ansprüchen genügt, und haben Interesse an einer honorarvergüteten Publikation?

Dann senden Sie bitte erste Informationen über sich und Ihre Arbeit per Email an *info@vdm-vsg.de*.

Sie erhalten kurzfristig unser Feedback!

VDM Verlagsservicegesellschaft mbH
Dudweiler Landstr. 99
D - 66123 Saarbrücken

Telefon +49 681 3720 174
Fax +49 681 3720 1749

www.vdm-vsg.de

Die VDM Verlagsservicegesellschaft mbH vertritt

Printed by Books on Demand GmbH, Norderstedt / Germany